Springer Biographies

The books published in the Springer Biographies tell of the life and work of scholars, innovators, and pioneers in all fields of learning and throughout the ages. Prominent scientists and philosophers will feature, but so too will lesser known personalities whose significant contributions deserve greater recognition and whose remarkable life stories will stir and motivate readers. Authored by historians and other academic writers, the volumes describe and analyse the main achievements of their subjects in manner accessible to nonspecialists, interweaving these with salient aspects of the protagonists' personal lives. Autobiographies and memoirs also fall into the scope of the series.

More information about this series at http://www.springer.com/series/13617

J. G. Weisend II • G. Terence Meaden

Going for Cold

A Biography of a Great Physicist, Kurt Mendelssohn

 Springer

J. G. Weisend II
Accelerator Division
European Spallation Source
Lund, Sweden

G. Terence Meaden
St Peter's College
University of Oxford, UK

ISSN 2365-0613   ISSN 2365-0621   (electronic)
Springer Biographies
ISBN 978-3-030-61201-6   ISBN 978-3-030-61199-6   (eBook)
https://doi.org/10.1007/978-3-030-61199-6

This Springer imprint is published by the registered company Springer Nature Switzerland AG
The registered company address is: Gewerbestrasse 11, 6330 Cham, Switzerland

Preface by G. Terence Meaden

It was a pleasure to be invited by Dr. John Weisend II to participate in this biography of Dr. Kurt Mendelssohn—a great Oxford physicist who died 40 years ago aged 74—because I am one of few surviving students or colleagues who knew him well. For me the friendship began from the point of view of 4 years as a doctoral student followed by 2 years as a post-doctoral fellow.

Before my first meeting with Dr. Mendelssohn, I had been an Oxford undergraduate at St. Peter's from 1954 to 1957 in the Department of Physics, beginning when the head of the Clarendon Laboratory was Professor Frederick A. Lindemann, also known as Lord Cherwell. He was another great man in the history of the Clarendon and the history of Britain by being Winston Churchill's close scientific advisor during the years of the Second World War.

Professor Lindemann's lecturing colleagues in the undergraduate teaching of low temperature physics included Kurt Mendelssohn and Nicholas Kurti. I much valued the physics lectures of all three. In this way, I became aware of the importance and excitement of the discoveries being made at Oxford about the properties of liquid helium and superconducting metals, and I soon viewed it as an ambitious aim to join one of the teams myself. Thus, it was that in my final year, I registered my name, as did many others, with Mr. T. C. Keeley—the managing director of the Clarendon— as a student who would be interested in undertaking post-graduate research in low temperature physics if the opportunity arose.

I took finals in June 1957, and next month was gratified to receive from Dr. Mendelssohn an invitation for interview regarding the possibility of embarking on a doctoral research project with him. His letter began: "I intend to start in the coming academic year two research projects in one of which you may be interested. The first concerns the investigation of the electric and thermal properties of the transuranic elements in metallic form. Plutonium and neptunium are now available and there is a hope that sufficient quantities of americium will be available in the not too distant future. This work will be coupled with the investigation of other heavy elements, in particular thorium and uranium".

A few days later, I was in Oxford for an interview, got accepted and was at once being shown the cryostat that the workshop had built to Dr. Mendelssohn's design for use in Oxford with thorium and uranium. He added that there would be another cryostat at the Atomic Energy Research Establishment at Harwell, where the work on the highly toxic radioactive metals plutonium and neptunium would be carried out using the protection of a glass-sided glove box. Senior scientific officer Dr. James A. Lee would be my supervisor at Harwell—another convivial lovely man and commendable first-rate scientist.

Thus, it was that in October 1957 I began getting to know Dr. Mendelssohn through frequent meetings with him and the staff of the technical and glass-blowing workshops. Working at first in Oxford, I took the cryostat and its supporting equipment through to the trial and functioning stages when studying the electrical properties of uranium down to liquid helium temperatures. At the same time, I began the first of many hundreds of car journeys to Harwell where I would test and operate a cryostat similar to what I had been using in Oxford but with the extra protection of an enclosed glove-box system. In this period while spending most of my time at Oxford, I came to know what an amiable likeable man Kurt Mendelssohn was, and progressively learnt what I needed to know cryogenically for the forthcoming research at Harwell using the new experimental systems.

The first visit to Harwell came at the start of October. Travelling in Dr. Mendelssohn's car allowed us to talk for nearly an hour during the journey, and it was much the same on the few subsequent trips that we made together because normally I would drive alone. I soon learnt what a brilliant all-rounder Kurt Mendelssohn was as a general conversationalist. KM (this being the usual way when speaking of him in the laboratory) was, like me, willing to discuss almost anything, having a kindly mindset and cheerful worldview. Born in 1906, he had followed an ambitious path in Germany from when he was young, and pursued it in his Oxford years to immense scientific advantage. His investigations had been pioneering and his timely changes of research direction in the 1930s proved highly profitable academically. He assured me that many scientists reach a peak of excellence in their late 20s and 30s, and I should bear that in mind. "Grasp the nettle", he said, means tackling the inevitable difficulties boldly, directly—much as he had done. This was an inspiring start to years of mutual respect and cooperation.

Weeks passed and experimental test runs interspersed with meaningful and successful runs came and went, but let us not forget the failures too. Often the runs that go wrong do not get into the thesis which can read as if every experiment had worked out well. The apparatus would spring a leak or vacuum pumps would break down or spot-welded wires would drop away in the face of the intense changes of temperature to which specimens were subjected. Dr. Mendelssohn was always an optimist. Although he took no part in any experimental runs with me, he followed my progress closely and cheerfully.

This was a world without personal computers and printed circuits. Calculations were made by 4-figure and 7-figure logarithm tables or the cranking by handle of a stepped-gear desktop Marchant or Monroe mechanical calculator the size of a typewriter. The laboratory seemed to be all Wheatstone Bridges, galvanometers, oscilloscopes, valves or electron tubes and other what-are-now-vintage electronic components, etc.

In Dr. Mendelssohn's group, the situation was that from some time before 1957 experimental research had been gathering a momentum of its own—as also in Dr. Kurti's group. Although the D.Phil. students each worked alone, we were colleagues too—forever obliging by helping, lending, repairing and so on. Dr. Mendelssohn forever sought progress reports and always accepted disappointments cheerily. I much valued how I had been selected for this project by a great physicist and busy man—and one who was a kindly father of five children with a caring, supportive wife Jutta.

Every year we would all be invited to a cocktail party at his splendid home at 235 Iffley Road which he had purchased new in 1937. There were usually some 30 of us crowded into the biggest room in his house, so well hosted by Jutta. There were other occasions too—dinner for four at his house—after I had married in France in 1965, when working at the Le Centre de Recherches sur les Basses Températures in the Faculty of Sciences at the University of Grenoble and brought my wife to visit him and my laboratory friends. Jutta was a charming hostess, perfect for an Oxford professor. Since 1955, KM had been an Oxford University Reader in Physics which is a position fully equivalent to full professor at any university elsewhere.

On one occasion, in my second year as I recall, I drove Dr. Mendelssohn to Harwell in my own car which was a white Triumph TR2 open-topped sports car. He loved it. It was only once though. Later, I learnt that he told members of the Harwell plutonium research team, "Terry brought me here in his sports car, and I kept saying to him 'seeing that this is a sports car why don't you drive much faster?'"

By the summer of 1960, KM judged that I had done enough research on plutonium and the other available actinide metals at Harwell and Oxford for my thesis although we had failed to find superconductivity in plutonium and neptunium despite reaching temperatures of 0.7 Kelvin by which temperature both thorium (at 1.37 K) and uranium (close to 0.75 K) had become superconducting. KM was urging me to write up my work, and he took on a new student, David Wigley, to whom I would teach the basics of applied cryogenics so that he would continue with the Harwell in-house research. Satisfied with my doctoral research, KM began discussing what examiners he might choose when the time would come in a few months. In the end, he selected faculty member Dr. Arthur H. Cooke of the Clarendon and Professor B. S. Chandrasekhar of Case Western University who in 1961 was on sabbatical leave in London.

My D.Phil. defence in July 1961 was satisfactory and uneventful. So KM was then saying that the work deserves to be in the Proceedings of the Royal Society, and no, he did not want to be a co-author either, which is something that few Ph.D. or D. Phil. supervisors anywhere would say. About then, he left for a conference and meetings in Tokyo and came back saying he had met a fine scientist who wanted to work in Oxford for a couple of years and learn English properly. He asked whether I would be interested with my doctorate out of the way. I agreed immediately. His name was Dr. Toyoichiro Shigi, Osaka University. Moreover, KM had located a commercial source that would supply a sufficient quantity of the rare helium-3 gas in order that at the Clarendon (not at Harwell) we may take specimens of plutonium and neptunium in protective sealed containers to lower temperatures than could be realized with helium-4. Temperatures approaching a third of a degree above absolute zero could be attainable. Toyo Shigi was another very agreeable colleague. I will mention an idiosyncrasy that was forever amusing. In our several laboratory rooms, each entrance door on its inside carried a blackboard on which anyone would write with chalk when discussing physics. Toyo took this a stage further. He and I would sit cross-legged on the wooden parquet floor and use the floor for chalking out our ideas as we discussed them. KM chuckled every time he found us thus engaged in this opportunistic expression of the Japanese work method.

Even more, I came to realize how KM was a polymath. Although he had spent so much of his life in pure and applied experimental physics, he was much interested in everything else too. When stopping off at Cairo one week early in 1965 on his way back to England from being a visiting professor at Kumasi University in Ghana, he started thinking in a fashion that only a scientist could about certain puzzles posed by the raising of the pyramids. He sought to understand the problems of the Third Dynasty Bent Pyramid and contemporary and later pyramids too. Pyramids were the study domain of specialist archaeologists known as Egyptologists who had had classical educations and not science-based ones. A scientist could refreshingly review the subject independently and dispassionately—and apply the conscientiousness of the scientific method too (viz. collect data, suggest hypotheses, then test them against sure facts in such an open fashion that anyone else repeating the exercise will obtain the same results). He was telling me about the Bent Pyramid when I was in Oxford on a couple of my return visits to see him in 1965 and 1966. This was the beginning of his 10 years research into problems involving structural factors and the reasons why pyramids were built at all, seeing that although they served as mausolea it was scarcely a sufficient motive to warrant the colossal long-term building efforts. How does one encourage so many thousands of men to labour long in undertakings about which most would never witness the final achievement? Surely there was something else too.

His delving into plausible explanations led to the 1974 publication of *The Riddle of the Pyramids*. But here I was in 1966 sympathizing with his approach partly because already as a physicist I had encountered my own problems when reflecting on the purpose of Stonehenge and seeking meaningful answers as to how priests and gang leaders could maintain willing cooperation over many decades from (in the case of Stonehenge) only farm workers of the local clans. Eventually, KM concluded that although the pyramids were built as cenotaphs "the object of the whole exercise was not the use to which the final product was to be put but its manufacture. What mattered was not the pyramid—it was *building* the pyramid". Because of my long-term interest in trying to understand Stonehenge I thought much the same and yet a little differently, viz. that the answer was also likely related to whatever was the core symbolism of the manifestly sacred pyramids. This could mean, for example, that, as was likely for the builders of Stonehenge, the heavenly rewards promoted by persuasive charismatic leaders were themselves stimulated by emboldened artful priests as to the high benefits in an afterlife for being compliant participants in the tough work of their present lives. Such can be the benefits when fostered powerfully by shrewd religious narrative and belief.

In the years from 1957 as I got to know KM better, he was nonetheless often absent attending national and international conferences in addition to board meetings arising from the founding of *Contemporary Physics* in 1959 and *Cryogenics* in 1960. He was also putting to good use visiting professorships combined with travel to Africa, the USA, Tokyo and other places and universities. He was a good speaker, a gifted writer. His *Quest for the Absolute Zero* published in 1966 is a marvel of clarity about an exciting difficult subject. Later, after visiting communist China in 1967 he summarized his findings of the tightly closed society by writing *In China Now,* published 1969.

In the early 1960s, he took on the task with Dr. K. D. Timmerhaus (University of Colorado) initiating a series of volumes called The International Cryogenics Monograph Series and invited me to write a book of my choice in low temperature physics. I chose to prepare *The Electrical Resistance of Metals* which came out in 1965 (Iliffe, London) and 1966 (Plenum, New York) and was reissued in 2013. I here repeat a relevant sentence from the preface to this book: "I am grateful to Dr. K. Mendelssohn, F.R.S., whose unrelenting enthusiasm has stimulated and developed my love for low-temperature physics". And again, in 1970 (although it did not come from KM) I received an invitation to prepare a physics review article for *Contemporary Physics*, which I gladly accepted, on the suggested subject of "conduction electron scattering and the resistance of the magnetic elements", published 1971.

In 1963 after 9 years in Oxford, I decided to move on and asked KM whether he would recommend me to Professor Louis Weil for a faculty position in the physics department of the University of Grenoble—and this he did with good grace although not really wanting me to leave the Clarendon right then. So I departed and became an assistant professor at a French university. Because there were useful studies still to do on the transuranic elements, research on the properties of such highly radioactive metals continued in England except that in the next decade everything would be done entirely at Harwell. In 1965 in Grenoble, KM visited me and Professor Weil. I was by then engaged to be married, so my fiancée and I took KM out to dinner at a choice Vietnamese restaurant. After Grenoble, I continued at Dalhousie University in Halifax (Canada) as an associate professor, where I set up new cryostats in the Clarendon mode, as so many others from Oxford had done at universities worldwide. I count myself lucky to have spent formative years with Dr. Mendelssohn because much of the Clarendon was a low temperature physics department raised to greatness, thanks to scientists like F. A. Lindemann, Kurt Mendelssohn, Nicholas Kurti, Sir Francis Simon, Arthur H. Cooke, John Daunt, D.K.C. MacDonald, Guy K. White, R. Bowers, J.L. Olsen, B.S. Chandrasekhar, Harry Rosenberg, Brebis Bleaney and Martin Wood together with many other researchers in pioneering aspects of advanced experimental and theoretical physics.

Bradford-on-Avon, UK   G. Terence Meaden
                                                                                       terencemeaden01@gmail.com

Preface by J. G. Weisend II

You have seen them on the road. Large tanker trucks, frequently white, bearing corporate names like Linde, Air Products or Air Liquide. If you look closely you may notice that they also say "Refrigerated Liquid Nitrogen" or, less frequently, "Liquid Helium". These trucks are a visible sign of the billion-dollar cryogenics industry.

Cryogenics—the study and use of extremely low temperatures—makes possible every MRI scan, the separation, collection and transport of gases like oxygen, nitrogen and argon for industrial and medical use, the flash freezing of food and the creation of immensely strong magnetic fields needed for the study of materials and of the very nature of matter itself.

Two of the largest scientific projects in the world today—the Large Hadron Collider (LHC) in Geneva, Switzerland, and the ITER fusion reactor currently under construction in France—are possible only due to superconducting magnets cooled by liquid helium. Spacecraft, cooled by superfluid helium, have tested Einstein's Theory of General Relativity and searched the cosmos in the infrared spectrum.

Cryogenics allows the efficient transport of liquefied natural gas from producers such as Algeria to users like Japan and the USA. The production and transport of liquid hydrogen, made possible by cryogenics, will be a vital part of any future hydrogen economy.

The temperatures involved are staggeringly cold. Helium gas becomes a liquid at 452° below zero Fahrenheit, i.e. $-269\ °C$. Superfluid helium at less than 456 °F below zero is colder even than interstellar space. Over time, humanity has learned how to create, keep and transport cold.

This is all quite new. Helium was not liquefied until 1908 and the first helium liquefier in the United Kingdom was not built until 1933. For a long time, cryogenics was strictly a small-scale laboratory science. The first helium liquefier in the United Kingdom was custom built, required the use of liquid hydrogen and only produced 20 mL of liquid helium. Today, the delivery of a million times that amount of liquid helium requires not much more than a telephone call and a credit card.

K.A.G. Mendelssohn played a major role in this development. An early refugee from the Nazis, he set up the first helium liquefier in the United Kingdom in 1933 and did fundamental research at Oxford University that increased our understanding of superconductivity and superfluid helium. He produced 188 scientific papers on such topics as the behaviour of superfluid helium films to the low temperature-specific heat of plutonium. He aided the transition of cryogenics from a small-scale science to a major industry by founding the leading scientific journal on cryogenics, setting up an ongoing international conference series, urging industrial development and training many students and visitors who went out and established their own research and industrial programmes.

Fascinated by travel and the world around him, Kurt Mendelssohn developed many collaborations between his group in Oxford and those of foreign countries, including countries such as China, India and Ghana who were just developing their industries. By the time of his death in 1980, Mendelssohn had met both Chairman Mao and Nehru and had personal contacts throughout the world.

Dr. Mendelssohn's efforts in developing collaborations, publications and conferences resulted in institutional structures that we use to this day in cryogenics.

Personally, Mendelssohn was generous, inquisitive, quick-witted with obvious joie de vivre. He was an excellent, meticulous, natural experimentalist. Moreover, as Professor Meaden's Preface makes clear, he was dedicated to both training his students and assisting them and his other collaborators with their research and careers whenever possible.

Kurt Mendelssohn had wide ranging interests and wrote books for the general public on topics as diverse as the history of cryogenics, his travels in China, the rise of Western science and the building of the pyramids. He contributed many articles to the popular media on physics, the teaching of science and the state of scientific research in the United Kingdom and other countries. He was frequently sought out by the press for interviews. His leisure time was taken up with photography, travel, gardening and the collecting of antique Chinese porcelain. He was frequently able to combine his work and leisure interests.

Unfortunately, Dr. Mendelssohn and his contributions are not as well-known as they should be, not even among current specialists in cryogenics. This book is an attempt to change this and make him better known to both specialists and to the general public. It is also a description of an inquiring, intellectual life well lived. It is not a complete story, but it emphasizes his contributions to science and his development of connections both within the cryogenic community and to the broader public.

When asked his interests by a journalist during a trip to India, Kurt Mendelssohn always an optimist, replied "anything that goes on in the world". It is in this spirit that his story is told.

Kurt Mendelssohn in 1967

Lund, Sweden   John G. Weisend II

Acknowledgements

Much of the pleasure that came from writing this book resulted from my interactions with the people who have assisted me over the years. Their help has been invaluable and I wish to express my deepest thanks to them. As always, any opinions or errors in the book are mine alone.

Dr. Monica and Dr. James Mendelssohn graciously answered questions about their father, his work and their life together. They have shared numerous family photographs and memories of their father. Two of Kurt Mendelssohn's colleagues at the University of Oxford, Nicholas Kurti and Ralph Scurlock, both provided memories of working with Dr. Mendelssohn and encouraged me in the writing of this book. John Vandore and Stephen Blundell helped connect me to a number of resources on Kurt Mendelssohn and Oxford. B. S. Chandrasekhar and Vinod Kumar Chopra shared their experiences of working in Dr. Mendelssohn's group at Oxford.

Terence Meaden was another of Kurt Mendelssohn's students. He contributed greatly to the volume by preparing the Preface, sharing personal experiences with me and carrying out several close readings of the manuscript. In working closely with Monica and James Mendelssohn, he was given numerous family and academic work-related photographs, and through several long interviews gained considerable personal family information about their father's life. Terence's participation and friendship have significantly improved the book.

B. S. Chandrasekhar and Steve Van Sciver both read drafts of the text and provided valuable comments and suggestions.

The staff at the University of Oxford Bodleian Library were very helpful in providing access to Dr. Mendelssohn's papers and in suggesting other sources of information.

Sam Harrison, Tom Spicer, Cindy Zitter and the entire team at Springer have all been invaluable in the production of this book.

This work has been carried out over a number of years during which time I have been on staff at the SLAC National Accelerator Laboratory, Michigan State

University, the European Spallation Source and Lund University. My colleagues in all these institutions have been very supportive of my efforts on this book.

My family, Shari, Rachel, Alex and Nick, have not only tolerated my absences due to this work but have also patiently listened while I explained the new and exciting item I had just learned. Their continuing support makes this book possible.

Contents

About the Authors

J. G. Weisend II is currently Deputy Head of Accelerator Projects and Group Leader for Specialized Technical Services at the European Spallation Source. He is also an Adjunct Professor at Lund University. He received his Ph.D. in Nuclear Engineering & Engineering Physics from the University of Wisconsin—Madison. He has worked at the SSC Laboratory, the Centre D'Etudes Nucleaires Grenoble, the Deutsches Elecktronen-Synchrotron Laboratory (DESY), the Stanford Linear Accelerator Laboratory (SLAC), the National Science Foundation and Michigan State University.

Dr. Weisend's interests include He II, large-scale accelerator cryogenics, cryogenic safety and the development of large international science projects. He is the Chairman of the Board of Directors of the Cryogenic Society of America (CSA). He has developed and taught classes in cryogenics at CSA short courses and webinars, at the US Particle Accelerator School, the European Cryogenics Course in Dresden and at Michigan State University and Lund University.

He is Chief Technical Editor of *Advances in Cryogenic Engineering* and a Section Editor-in-Chief for the *European Physical Journal/Techniques & Instrumentation.* He is one of the series editors for the International Cryogenic Monograph Series.

In addition to co-authoring more than 70 technical papers, Dr. Weisend is the author of *He is for Helium*, the co-author (with N. Filina) of *Cryogenic Two-Phase Flow*, co-author (with T. Peterson) of *Cryogenic Safety* and the editor of the *Handbook of Cryogenic Engineering* and of *Cryostat Design*. He writes a regular column "Cryo Bios" for the publication *Cold Facts* and is a member of both the Cryogenic Engineering Conference and International Cryogenic Engineering Conference Boards.

G. Terence Meaden is a life-time academic. From 1954 to 1957 he was an undergraduate in the Department of Physics at the Clarendon Laboratory, Oxford, and in 1957 was invited by Dr. Mendelssohn to join his low temperature physics research team. Doctoral research involved working at both the Clarendon and the Atomic Energy Research Establishment at Harwell. Afterwards, Dr. Meaden remained at Oxford as a post-doctoral fellow until 1963 when he left for a faculty position (Assistant Professor) at Grenoble University and the Centre de Recherches sur les Très Basses Températures.

He was next a tenured Associate Professor of Physics at Dalhousie University, Halifax, Canada, for several years before retiring early for family and health reasons and returning to Europe. In Oxford, as a Fellow of the Royal Meteorological Society, he formed the Tornado and Storm Research Organisation and launched *The Journal of Meteorology* which he edited for 27 years. He was also a consultant to the New-Build Nuclear Power Industry. A 30-year research study into megalithic aspects of the prehistory of Britain and Ireland is ongoing, and in 2001–2009 an Oxford University M.Sc. degree in archaeology was obtained. He has researched and written books on physics, meteorology, Neolithic and Bronze Age archaeology, archaeoastronomy, rock art, Stonehenge, empirical philosophy, and the history of a French abbey—besides 200 papers and encyclopaedia and other book contributions.

Chapter 1
Foundations: Berlin 1906–1925

*"So this was Berlin—the city where I was born, had gone to
school and studied"*
The World of Walter Nernst *K. Mendelssohn, 1973*

Kurt Alfred Georg Mendelssohn was born on Sunday the 7th of January, 1906 in the
Schoenberg suburb of Berlin to a family with a long history of intellectual achievement.
He was the only child of Ernst Mendelssohn (1873–1948) and Elizabeth (*née* Ruprecht
1879–1952). The family usage of the name Mendelssohn (Mendel's sohn) was chosen
by Moses Mendelssohn (1729–1786) the son of Mendel Heyman of Dessau.

Mendel Heyman came from a quite humble origin and was an excellent scribe
who married Bela Rachel Sara Wahl, a descendant of the great Jewish scholar Moses
Isserles founder of a Yeshiva in Krakow and world-renowned scholar. Moses was a
German Jewish philosopher, critic, Bible translator and commentator who greatly
contributed to the efforts of Jews to assimilate into the German bourgeoisie. He was
proficient in German and Hebrew and also learned Latin, English, Greek and French.
Throughout his life, Moses showed a voracious desire for new knowledge and had
very wide interests. These are traits that his descendent Kurt would share.

In 1750, Moses moved to Berlin to be a tutor in the house of a Jewish silk
merchant. He strongly believed that Jews, while remaining true to their faith, should
involve themselves in the outside world. He championed the idea that Jews should
be full participating citizens of the countries in which they lived. Moses was an
advocate of the German language and wrote with great style. He produced one of the
first translations of the Old Testament into German. "Within one generation, the
Mendelssohn Bible was in the home of nearly every literate Jew in Western and
Central Europe" [1]. In his philosophical works, Moses Mendelssohn attempted to
combine reason and faith. He believed that one could use reason to prove things such
as the existence of the soul and of God.

The original version of this chapter was revised. The correction to this chapter is available at
https://doi.org/10.1007/978-3-030-61199-6_11

J. G. Weisend II, G. T. Meaden, *Going for Cold*, Springer Biographies,
https://doi.org/10.1007/978-3-030-61199-6_1

Fig. 1.1 Philibert
Mendelssohn who was
grandfather to Kurt
(Courtesy:
M. Mendelssohn)

Later, we will see that Kurt Mendelssohn also had a wide range of interests and a tendency to look at the big picture. Kurt did not try to combine religion with science. Kurt would probably also have enjoyed Moses' supposed response to a German customs officer's question of "In what merchandise do you trade Jew?" His answer "In reason sir, a commodity with which you have no acquaintance" [2].

Moses' grandson was the renowned composer Felix Mendelssohn (1809–1847), and Kurt was a descendant of Moses' younger brother Saul (1773). The strong intellectual tradition of the family continued and Kurt's paternal grandfather Philibert Mendelssohn (1830–1915) the great grandson of Saul, was a mathematician who had the position of 'Koenigliche Rechnungsrat' in the Prussian State Survey, a rare distinction for a Jew at that time. Philibert's father was a Jewish teacher and one of Kurt's great grandfathers on his mother's side was a Lutheran pastor, but despite this background Kurt and indeed his parents, were not religious.

Philibert had a clear plan for the future of his three sons which was expressed as 'I have paid for your school education and now it is time for you to become self-supporting' (Fig. 1.1).

Although none of Philibert's sons were professional scientists his three grandsons became successful academic scientists. Kurt and his cousin Franz (later Sir Francis) Simon became physicists and Kurt's other cousin, Heinrich Mendelssohn, was a zoologist and one of the founding fathers of Tel Aviv University. Kurt's father, Ernst, was a foreign representative for manufacturers and wholesalers of men's wear, and the family were tolerably well off when Kurt was born in 1906 (Fig. 1.2).

Ernst had inherited the Mendelssohn enquiring mind and he spent all his spare time reading over a range of subjects, literary classics, politics and popular scientific books and magazines, mainly on nature study (Figs. 1.3, 1.4, and 1.5).

This enthusiastic reading habit resulted in a capacious knowledge with which he was able to inspire Kurt's natural inquisitiveness as well as providing a large library of books for him to read when he grew older. From an early age Kurt's father took him on Sunday walks and introduced him to the exciting world of science and technology, and, interleaved with sections from the Odyssey, there were visits to

Fig. 1.2 Ernst and
Elizabeth Mendelssohn,
parents of Kurt (Courtesy
M. Mendelssohn)

Fig. 1.3 Kurt age
2 (Courtesy
M. Mendelssohn)

museums, exhibitions and lectures on popular science [3]. One of the main museums
of interest was the Egyptology Museum. Egyptology was extremely popular at this
time. Kurt's research and book, 'The Riddle of the Pyramids' echoes these early
experiences. Like his father, Kurt early became a passionate reader.

Kurt's interest in science was recognized by his father and when he subscribed to
a scientific magazine for himself a junior version was provided for Kurt. The Sunday
walks were a major inspiration for Kurt's insatiable quest to find out about, understand and analyze as much information about the world that he lived in. While still a
very young child, his father would sometimes take him to visit Philibert and while

Fig. 1.4 Kurt age 6 (Courtesy M. Mendelssohn)

Fig. 1.5 Kurt after recovering from encephalitis (Courtesy M. Mendelssohn)

the adults played cards the smaller children entertained themselves. On one occasion an older cousin had declared that Philibert's caged birds were finches, the noticeably young Kurt then explained that in fact they were Javanese Rice birds!

Kurt's mother was a keen gardener which led Kurt to his own pleasure in gardening. She also took him to concerts which introduced him to the world of classical music which he enjoyed, especially Wagner being played very loud.

This comfortable world was seriously disrupted in 1916 when Kurt's father was called up to join the army. This put Kurt's mother under so much emotional stress that she was unable to look after him properly. The family were living on the top floor of a block of flats and Kurt (1919) embarked on a new project and asked his mother not to go into his room. The frequent sight of Kurt leaving his room covered in dust carrying a dusty bag of unknown contents, was more than her curiosity could resist. Entering his room, she discovered, to her horror, that he had been cutting a hole in the ceiling to be extended to go through the roof so that Kurt could put his telescope out to watch the stars.

There were several unsupervised boys who took to wandering around Berlin, being exposed to a harsh and cruel city while Germany ultimately lost WW1. Kurt considered this a valuable preparation for his future life and, blended with the bitter message of The Threepenny Opera (Weil and Brecht), formed the basis of his attitude of human interaction. The ignominy of losing the war and the appalling post-war economic downfall, left the Germans in a very unstable situation.

When Kurt was about 12 years old, he fell ill of encephalitic lethargica and possibly post-encephalitic Parkinsonism. The author Constantin von Economo in his 1917 paper "Die Encephalitis Lethargica", explained how the disease was characterized by high fever, ophthalmoplegia, mental confusion, and lethargy. This epidemic coincided with the height of the great influenza pandemic in 1918 and lasted some 10 years.

Between the ages of 6 and 19 Kurt attended the Goethe Schule in Berlin, where he revealed his singular trait for concentrated hard work on subjects which interested him and a less diligent application to those that did not [3]. Work was graded on a scale of 1 (high) to 5 (low). Favored subjects were Physics, Earth Sciences, Drawing and Gymnastics always gaining grades of 1 or 2. In subjects of secondary interest to him, mathematics and most of the arts, he would get mediocre grades of 3 or 4. Kurt's mother, who was very ambitious for him, regularly admonished him, in no uncertain terms, for his poor grades which she knew he could improve with a bit more effort. No matter how fierce the dressing down, it made no more impression on him than water off a duck's back and he continued to concentrate on subjects that intrigued him.

Berlin, in 1906 when Kurt was born, was at its peak as the capital of the newly united Germany. Kaiser Wilhelm II desired to make Berlin into another Paris or London. While he was arguably less than successful in this, Berlin was quickly becoming much more cosmopolitan. A new research institute, the Kaiser Wilhelm Gesellschaft, was started in 1909. The collections of Berlin art museums were greatly expanded by the addition of paintings by Botticelli, Titian, Rembrandt, Raphael, Durer, and others. Archeological museums, in particular, expanded.

Excavations at Tell al-Amara, Egypt between 1911 and 1914 brought large numbers of artifacts to the Berlin Egyptian Museum including the famous bust of Nefertiti. Large scale artifacts from the Near East including the Pergamon Altar, the Ishtar gate of Babylon and the Market Gate of Miletus as well as the gold of Troy all arrived in Berlin in the early twentieth century. Public lectures on science, technology and the arts were (and still are) an important part of German social life. All these attractions were available to the curious public.

This early exposure to science, history and foreign cultures had a major impact on Kurt Mendelssohn's later life. It is not surprising, for example, that he would develop a keen interest in Egyptology. There was also a subtler effect. Throughout his career, Kurt Mendelssohn put a great deal of emphasis on making science understandable to the general public. He did this through his books, and through public lectures and appearances in the media. His childhood had shown him that educated laymen, such as his father, were truly interested in learning about the latest scientific advances.

As previously mentioned, Kurt Mendelssohn attended the Goethe Schule. His differing performances in drawing and mathematics are echoed in his later life. Most of his adult hobbies, such as photography or the collecting of oriental ceramics, were visual in nature. He was also very particular about the layout and design of the books he wrote. While Mendelssohn certainly used mathematics professionally, he frequently displayed experimental results and conclusions best through drawings and data plots. In fact, as we will see, he was quite innovative in the use of graphics to explain scientific results. This helped make his work more accessible to both scientists and the broader public.

References

1. G. R., Marek, *Gentle Genius: The Story of Felix Mendelssohn*, (Funk & Wagnells, New York, 1972).
2. C., Potak, *Wanderings*, (Fawcett, New York, 1979).
3. D., Shoenberg, "Kurt Alfred Georg Mendelssohn", Biographical Memoirs of the Fellows of the Royal Society, Vol. 29 (1983).

Chapter 2
University: Berlin 1925–1931

"It was a moment of extreme elation when, plotting the first graph, I saw this new evidence and suddenly the strain of innumerable nights spent in the laboratory had vanished into nothing"
The World of Walter Nernst *K. Mendelssohn, 1973*

In 1925, Kurt Mendelssohn entered the University of Berlin.[1] He had at one point thought about being a stockbroker. However, at the university he studied physics, mathematics, chemistry and psychology. He took great pride in having studied under Einstein, Planck, Schrödinger and other leaders of the new "Modern Physics".

He started research at the University's Physikalisch Chemisches Institut (Institute for Physical Chemistry) in 1927 studying under his cousin Franz Simon (Fig. 2.1) who was 12 years older. Franz Simon had done his dissertation at Berlin under Walter Nernst who formulated the third law of thermodynamics. Simon, Nernst and another of Nernst's students, F. A. Lindemann, would have a profound influence on Mendelssohn's life.

In order to understand Kurt Mendelssohn's work at Berlin and its impact on his later activities, a few sentences on the third law of thermodynamics are in order. Nernst's development of the third law was motivated by the problem of predicting chemical reactions. Specifically, did a given set of chemical components react spontaneously, did the reaction require the addition of energy or were the components in chemical equilibrium? It was known that the important parameter was the difference in the free energies of the reactants and products of the reaction. For a reaction at constant temperature and pressure, if the free energy of the products was less than that of the reactants the reaction would be spontaneous. If the free energy of the products was more than that of the reactants, energy would have to be added for

---

The original version of this chapter was revised. The correction to this chapter is available at https://doi.org/10.1007/978-3-030-61199-6_11

[1] After World War II this University, located in East Berlin, was renamed the Alexander Humboldt University.

© The Author(s), under exclusive license to Springer Nature Switzerland AG 2021,
Corrected Publication 2021
J. G. Weisend II, G. T. Meaden, *Going for Cold*, Springer Biographies,
https://doi.org/10.1007/978-3-030-61199-6_2

Fig. 2.1 Franz Simon
"Franz Eugen Simon",
*Biographical Memoirs of
Fellows of the Royal
Society,* Volume 4 (1958)

the reaction to take place. If the free energies of the reactants and products were equal then the system was in chemical equilibrium.

This is simple enough, but the problem was that while the total energy of a substance could be measured there was no obvious way to determine the free energy from the total energy. Nernst postulated that at a temperature of absolute zero, the difference between the free energy and total energy of a substance must be zero and that the difference must approach zero as absolute zero temperature is approached. With this hypothesis, it was now possible to calculate the free energy of a substance as a function of temperature starting at near absolute zero temperature. This required knowing the specific heat and thermal expansivity of the substance (i.e. the temperature rise and expansion due to a known heat input) as a function of temperature. Thus, to prove his theory, Nernst and his group had to reach very low temperatures and make precision measurements of specific heat and thermal expansivity. Over a number of years, Nernst aided by Frederick A. Lindemann, his brother Charles Lindemann, Franz Simon and others made measurements that supported Nernst's hypothesis. As a result of their efforts, the third law of thermodynamics is accepted today as one of the fundamental laws of nature.

However, as is frequently the case in science, the nature of the question changed. Nernst developed the third law to allow the prediction of chemical reactions and his result was used successfully for this purpose. But an alternative way of expressing the third law said that as absolute zero temperature is approached, the entropy[2] goes to zero and any entropy difference also goes to zero. The third law really says fundamental things about matter at absolute zero. At a temperature of absolute zero, the entropy of any substance is zero (this represents the zero point energy). Furthermore, since any cooling process requires an entropy difference, the third law says

---

[2]The entropy is a thermodynamic quantity best thought of as representing the degree of disorder in a system.

that while you may get infinitesimally close the absolute zero it is impossible to reach it.[3]

The absolute temperature scale used in science has the unit Kelvin (K) 0 K = −273 °C or − 459 °F. Helium liquefies at 4.2 K at atmospheric pressure whereas room temperature is about 300 K. Cryogenics, the study of materials and phenomena at low temperatures, is generally defined as concerning itself with processes below 120 K. Nernst's group became interested not in chemistry but in the study of materials at very low temperatures, that is, cryogenics.

After completing his doctorate in 1921, Simon remained at the Physikalisch Chemisches Institut where he continued to work on problems in cryogenics and thermodynamics. He organized a cryogenics group and had a hydrogen liquefier and a helium liquefier built. His laboratory was the fourth in the world to liquefy helium, the others being in Leiden, Toronto and the Physikalisch Technische Reichsanstalt in Berlin.

Graduate school is designed to teach how to conduct research as much as it is designed to teach a given specialty. Kurt Mendelssohn described the process: "Research, like any other job, has, of course, to be learned. Even the most gifted and successful scientist knows that the sparks of genius and the moments of inspiration are few and far between. The rest of his life will be devoted to patient work, careful planning and perfection of those techniques which he has to master for the exploration of his chosen field" [1].

Kurt Mendelssohn's work in Simon's group was centered in two areas, specific heat measurements at cryogenic temperatures and the construction of small helium liquefiers. The first of these led to his doctorate, the second led to Oxford.

Kurt Mendelssohn's thesis project was the measurement of the specific heat of solid hydrogen. It was known at this time that hydrogen existed in two states, parahydrogen in which the nuclear spins are oriented antiparallel to each other, and orthohydrogen in which the nuclear spins are oriented parallel to each other. The Pauli exclusion principle—part of the radical new theory of quantum mechanics—predicted that the energy of orthohydrogen at absolute zero should be higher than that of parahydrogen. Consequently, the specific heat of orthohydrogen should rise as the temperature approaches absolute zero before a rise is seen in the specific heat of parahydrogen. This is the effect that Mendelssohn was trying to measure. This was a very difficult experiment, due to the low temperatures and small effects involved. However, Mendelssohn was finally able to measure a rise in the specific heat of orthohydrogen at 5 K while the specific heat of parahydrogen continued to decrease with temperature down to 3 K, the lowest temperature reached in the experiment. The results were consistent with the predictions of quantum mechanics and thrilled Mendelssohn as he described them in the quotation at the start of this chapter.

The results also cleared up a potential problem with Nernst's third law of thermodynamics. Earlier measurements of the chemical potential of hydrogen ran

---

[3]The current world record is about half a nanoKelvin (5×10^{-10} K).

Fig. 2.2 Five Nobel Laureates Together: Nernst, Einstein, Planck, Millikan and von Laue at von Laue's house in Berlin, 1923 [1]

counter to that predicted by the third law. Simon had theorized that the problem was not with Nernst's law but rather due to exactly the type of specific heat rise that Mendelssohn had measured, which was why Simon had given him this as his thesis topic.

That a single thesis project could provide support to both the third law of thermodynamics and to quantum mechanics illustrates the intellectual atmosphere of the time. Experimental and theoretical work was developing the concepts of quantum mechanics and relativity which fundamentally changed our understanding of the universe. This work was being done by a small group of scientists, known to each other and principally located in Northern Europe. An illustration of the connections between scientists at this time is shown in Fig. 2.2.

In his 1973 book, *The World of Walther Nernst,* Kurt Mendelssohn portrays the rapid changes of which he was a part (Fig. 2.3).

Mendelssohn's thesis experiment operated at temperatures as low as 3 K. These temperatures required the use of liquid helium, which was not easy to come by at that time. Early on, Simon had decided that rather than build a single large helium liquefier that continuously makes liquid helium for the whole laboratory he would develop miniature liquefiers that produced just enough liquid helium for each individual experiment. In effect, each experiment would have its own liquefier. This approach had three advantages. First, there would be no need for a complicated central helium facility that had to be maintained. Second, it minimized the amount of

Fig. 2.3 Kurt Mendelssohn while working on his thesis (Courtesy of M. Mendelssohn)

hard-to-get helium gas required for each experiment. Lastly, since each student would have to build his own liquefier they would develop a detailed practical understanding of the liquefaction of helium. "Simon set great store by this arrangement, since he regarded skill in cryogenic techniques as an essential part of the low temperature physicist's qualifications" [2].

The first miniature liquefiers built by Simon used the desorption method. In this technique, helium gas was cooled to roughly 20 K by heat exchange with a liquid hydrogen bath. The cold gas was then forced under pressure to absorb on to an activated charcoal layer. When the gas pressure is reduced, helium starts to desorb from the charcoal. However, the helium that remains absorbed on the charcoal is cooled to less than 5 K in this process. By next raising the helium pressure to slightly less than two atmospheres, the helium liquefies. An advantage of this technique does not require the use of high pressures. This type of liquefier was integral to Mendelssohn's thesis experiment.

In 1930, Kurt Mendelssohn was awarded his Ph.D. in physics and decided to stay at the university to work as Simon's assistant.

References

1. K.A.G. Mendelssohn, *The World of Walter Nernst*, (University of Pittsburg Press 1973).
2. N. Kurti, "Franz Eugen Simon", Biographical Memoirs of Fellows the Royal Society, Volume 4 (1958).

Chapter 3
Transitions: Breslau/Oxford 1931–1933

> *(Dr Mendelssohn) "placed all his knowledge and experience*
> *unreservedly at the disposal of the department. But for this, it*
> *would have scarcely been possible to obtain without hitch or*
> *trouble, liquid helium within 1 week of the arrival of the*
> *apparatus in Oxford"*
> Announcement of the first helium liquefaction in England
> *Lindemann and Keeley,* Nature, *February 11, 1933*

In 1931 Franz Simon accepted the position of Professor of Physical Chemistry at the Technische Hochschule in Breslau, Germany (currently Wroclaw, Poland). He was hesitant to leave Berlin but he recognized it as an opportunity to start his own research program. He invited Kurt Mendelssohn, who had been working as his principal assistant for the past 2 years and Nicholas Kurti (Fig. 3.1), who had just finished his doctorate under Simon, to come with him to Breslau. While Technische Hochschule literally translates as "Technical High School", under the German system such institutions are well regarded universities specializing in technical subjects. The American equivalent would be schools such as the Massachusetts Institute of Technology or Caltech. Among Simon's students at Breslau were Heinz London from Germany and Rosteslav Kaichev from Bulgaria.

Franz Simon was invited by G. N. Lewis of the University of California at Berkeley, Chemistry Department to spend the spring semester of 1932 as a visiting professor at Berkeley. The major result of his stay at Berkeley was the development of another type of miniature liquefier, the expansion liquefier. In this method, helium gas is compressed to high pressures and cooled via a bath of liquid hydrogen to 20 K. The pressure in the liquid hydrogen bath is reduced using a vacuum pump. As the hydrogen is pumped out, the temperature drops and any remaining hydrogen freezes and sublimes (converts directly to vapor). Thus, at the end of this step, the container of helium gas is thermally isolated from the rest of the system and at a temperature of about 10 K. Next the high pressure (150 atmospheres) helium gas is expanded

The original version of this chapter was revised. The correction to this chapter is available at https://doi.org/10.1007/978-3-030-61199-6_11

J. G. Weisend II, G. T. Meaden, *Going for Cold*, Springer Biographies, https://doi.org/10.1007/978-3-030-61199-6_3

Fig. 3.1 Nicholas Kurti in
1926. "Nicholas Kurti,
CBE.", Biographical
Memoirs of Fellows of the
Royal Society, Vol.
46 (2000)

through a room temperature valve. As the pressure drops in the container, the remaining gas expands, cools and liquefies, resulting in the container being between 80% and 100% full of liquid helium. In order for this method to work, the specific heat of the metal container must be very low and the temperature of the helium gas must be less than 43 K so that the gas cools as it expands; thus, the requirement for cooling with liquid hydrogen. A cross section of the expansion liquefier is shown in Fig. 3.2. Like the desorption liquefier before it, this technique requires very little helium gas, it retains all the gas used and it can be built directly into experimental systems. The expansion liquefier differs from the desorption liquefier in that it does require high pressure helium gas. However, by this time, equipment capable of compressing helium gas to these pressures was available.

Meanwhile, back in Breslau, Kurt Mendelssohn was looking after matters in the institute.

Franz Simon recognized early on the danger of the rise of Nazism and as a precaution had left his wife and two daughters in Switzerland while he went to California. As such, the Simon house in Breslau was free and Simon let Mendelssohn, Kurti, London and Kaichev to live there during his absence. The four physicists got along quite well and worked hard together to set up a new research facility. They had nicknames for each other. Kurti was 'Niko' and Kaichev was 'Ro'. London had earned the nickname 'Behemoth' apparently due to his clumsiness. Mendelssohn, being the senior of the four, was known as Onkel (Uncle). While he collaborated with others throughout his life, it is really only during these periods in Berlin and Breslau that one gets an impression of him working as a member of a group of peers (Fig. 3.3).

The need to create a laboratory from scratch, the absence of Simon and increasing political turmoil prevented a lot of research work from being done during Mendelssohn's stay in Breslau. The group continued to develop Simon's expansion liquefier; and London started work on losses in superconductors operating at radio frequencies. Mendelssohn became interested in searching for the magneto–caloric effect in tin. The result of this effect meant that if you had a superconducting piece of tin thermally isolated and caused it to transition from the superconducting state to a

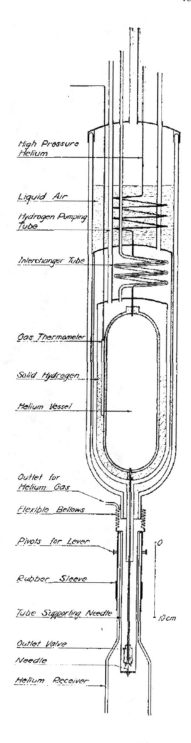

Fig. 3.2 Cross Section of Simon Expansion Liquefier. (Reproduced from A. H. Cooke, B. V. Rollin, and F. Simon, "A New Form of Expansion Liquefier for Helium", *Review of Scientific Instruments* 10, 251 (1939), with the permission of AIP Publishing)

Fig. 3.3 Kurt Mendelssohn and colleagues in Breslau circa 1931–1933. Top Left—Nicholas Kurti, Middle Left—Rostislav Kaichev, Bottom Row, 3rd and 4th from left—Barbara Zarniko, Kurt Mendelssohn (Courtesy J. Mendelssohn)

normally conducting state by applying a strong magnetic field, the tin sample would become colder. This would result in a new way of providing cryogenic cooling. Due to his short stay in Breslau, Mendelssohn did not have time to look for this effect experimentally. As will be seen, he returned to this work once he moved to Oxford.

In 1919, Frederick Lindemann, who had studied under Walther Nernst in Berlin, was appointed to be the Dr. Lee's Professor of Experimental Philosophy at the University of Oxford. This position made him the head of the Clarendon Laboratory and responsible for the physics program at Oxford. At this time there was effectively no physics research program. The Clarendon Laboratory had no research staff, no electrical power, and any instruments that it did own were still packed in boxes. Lindemann took on developing a viable physics program as his principal task. As a result of his experience in Berlin with Nernst, one of the first research areas that

Lindemann started was cryogenics. He purchased a hydrogen liquefier from Simon's laboratory in Berlin and started work on a small scale. This first hydrogen liquefier never worked properly. After Simon's group moved to Breslau, Lindemann bought an improved hydrogen liquefier which became a mainstay of the Clarendon Laboratory.

In 1932, Lindemann invited Mendelssohn to come to Oxford for a year to assist them in their low temperature research. Lindemann was now quite interested in moving on to work at liquid helium temperatures. Mendelssohn's proposal was to work in two areas in Lindemann's laboratory. First, the physical properties of superconductors (particularly the impact of crystal structure on superconductivity) and second, the gas desorption from copper surfaces at very low temperatures. Lindemann agreed with these research topics and was very interested in getting Mendelssohn to come to Oxford under a Rockefeller Foundation grant. Setting up this grant took time and by the end of 1932 there was still no resolution.

In the meantime, Frederick Lindemann wished to purchase a small helium liquefier (based on the expansion process) from Simon's group and have Mendelssohn come over to Oxford to set it up. The timing of this was not just driven by Lindemann's desire to conduct experiments with liquid helium. Oxford's traditional rival, Cambridge University, had received a large sum of money from the Mond family to build a cryogenics laboratory to be run by the Russian physicist Peter Kapitsa. This laboratory would include a helium liquefier and would then be the first place to liquefy helium in the United Kingdom. By bringing Mendelssohn over with a helium liquefier, Lindemann would thus show that Oxford did it first. Kurt was completely unaware of this aspect of his visit to Oxford and only realized the truth when he had a rather cool reception from Kapitsa and all was revealed. The academic year had already started in Breslau. Kurt Mendelssohn agreed to visit Oxford in early January 1933, during the Christmas break, to set up and operate the liquefier. In preparation for this, he corresponded in some detail with Lindemann's assistant, T. C. Keeley, to ensure that the appropriate equipment and facilities were available. Items such as vacuum-insulated flasks and valves that were not available in Oxford were acquired by Mendelssohn in Germany and brought to England. This degree of thoroughness and preparation was typical of his approach to both work and hobbies. He did not tend to do things half way. During what must have been a fairly hectic month, Kurt Mendelssohn and Jutta Zarniko were married on December 19, 1932.

This is how Kurt and Jutta got to know each other, as recounted by Kurt Mendelssohn's daughter Monica Mendelssohn in 2020.

"Kurt had joined Franz Simon's group in 1927 and while everybody was focussing on the science, a set of romantic liaisons came about which involved all three of the Zarniko sisters. Martin Ruhemann was a senior member of the group and was involved with Barbara Zarniko, one of Simon's earliest students, who was in the same year as Kurt who was also attracted to Barbara but the presence of Martin did not let that develop. Meanwhile, Jutta who was 7 years younger than Barbara was at boarding school in Berlin but went home to her family over the Christmas holidays. New Year's Eve is a huge celebration in Germany and Barbara and Martin wondered who they should invite to accompany Jutta for the festivities and decided on Kurt.

Fig. 3.4 Fransiska, Jutta,
Barbara, Jim, Kurt, Martin,
Liza with Corinna on her
lap, and Stephen (Courtesy
M. Mendelssohn)

On New Year's Eve 1929 Kurt and Jutta's introduction to each other led to their wedding in December 1932, which lasted for nearly 50 years until my father's death in 1980.

Barbara and Martin moved to the Kharkov in the Ukraine where Barbara gave birth to their son Stephen. To help with this event Fransiska, the middle Zarniko daughter, was staying with the Ruhemanns in Kharkov. During her stay Jim Crowther visited the Ruhemanns and another light-bulb moment occurred between Jim and Fransiska resulting in their marriage once Jim had arranged the transit from Germany to England for Fransiska. Jim Crowther was the doyen of scientific journalism and the author of many books on scientific topics.

In the summer of 1938, all three couples were settled in England and Liza Zarniko came over from Germany to celebrate her sixtieth birthday with her married daughters. The Zarnikos were an old Prussian family whose ancestors had been sent off with the Teutonic Knights to Christianise the Pagan Prussians.

Figure 3.4 is a photograph, taken in the back garden of 235 Iffley Road, and was the last that any of the family saw of Liza as she was in East Germany at the end of WWII. Very shortly before she died, she was moved to Aachen where Jutta went to visit her, but Liza's memory was so impaired that she could not remember who Jutta was."

Dr. Mendelssohn arrived in Oxford the first week of January 1933 and was at once "going for cold". On January 13 the first liquid helium in England was produced.[1] It should be pointed out that setting up a new apparatus in an unfamiliar laboratory and getting it to work properly in 1 week is a real testament to Mendelssohn's skills particularly since he was suffering from a bad cold at the time. Lindemann was certainly impressed and quite pleased that Oxford had accomplished this feat before their rivals at Cambridge. As an added tweak to Cambridge's pride, Frederick Lindemann arranged for the announcement of the helium liquefaction to be published in the same issue (February 11, 1933) of the journal *Nature* as the story about the recently dedicated but still empty Mond Laboratory. He also wrote yet another letter to the Rockefeller Foundation asking for funding to support Kurt Mendelssohn's work in Oxford. It looked likely that such a grant would be

---

[1]The room in the Clarendon in which the liquefaction was accomplished later became a Men's Room which led campus wags to comment that the commemorative plaque should state "that here K. Mendelssohn first passed liquid helium in England". KM probably appreciated the joke.

Fig. 3.5 Parade of Reichsbanner members. Kurt Mendelssohn is second from left in the first row carrying the Schwarz Rot Gold banner. Ernst Mendelssohn is fourth from left in the first row (Courtesy of Monica Mendelssohn)

given and that Kurt Mendelssohn would go to Oxford in October of 1933, but events in Germany moved more quickly than that.

Kurt Mendelssohn had been in Breslau since 1930. Breslau became a well-known hot spot for the development of Nazi terrorism and people were genuinely scared for their safety. In Breslau Heines was the Nazi Police Chief. Hitler became Chancellor in January 1933 and in the middle of March 1933 he declared that the only legal political party was the Nazi party and that members of any other type of political party were to be rounded up and dealt with to free Germany from non-Nazis.

Kurt Mendelssohn was now extremely frightened, as in his late teens Kurt joined his father in becoming a member of the Reichsbanner Schwarz-Rot-Gold which was an organization in Germany during the Weimar Republic. It was formed in February 1924 by members of the Social Democratic Party of Germany, the German Centre Party, and the German Democratic Party. Its goal was to defend parliamentary democracy against internal subversion and extremism from the left and right, to teach the population to respect the new Republic, to honor its flag and the constitution. Its name is derived from the Flag of Germany adopted in 1919 (see Fig. 3.5). Kurt Mendelssohn's membership meant that he certainly would be a wanted man if he fell afoul of the brown shirts.

He had another two points against him. While baptized a Lutheran, he had a very famous Jewish background. He was an intellectual, and worse, a physicist. Hitler believed that Einstein's Theory of Relativity was an example of degenerate "Jewish Physics".

Monica Mendelssohn explains what happened next: "At Easter, in April 1933, Kurt and Jutta went to Berlin where Jutta was going to stay with her mother in central Berlin, Kurt then left to go to Baabelsberg in the west of Berlin to visit his parents. Somewhere along his journey he felt someone punching him in the back and turned round to come face to face with a brown shirt who aided by other brown shirts was tying a cordon of rope around a group of people being taken off to be questioned about their political affiliation. Kurt was completely terrified by being so nearly captured and aware that he had to get out of Germany immediately. He phoned Jutta to tell her that he was leaving for England the next day and that she should pack up and follow him. Jutta was in no danger herself as she had no political commitment and was not Jewish." His destination was Oxford to seek a position with Lindemann.

Like Mendelssohn, Lindemann recognized early on the danger of the Nazis. He realized that there would be a flood of refugee scientists and saw this as an opportunity to bolster the research program at the Clarendon Laboratory. The problem was money. Oxford University did not have sufficient resources to create a lot of new permanent positions. The Rockefeller Foundation, whose grant money was originally envisioned to support Mendelssohn's visit, also was not the solution. Their rules required that the grant recipient return to his home country at the end of the grant, something that the German refugee scientists like Mendelssohn would be unable to do. Lindemann solved this problem by going to Sir Harry McGowan who was an old friend and the Chairman of Imperial Chemical Industries. Lindemann convinced McGowan that it would benefit England, ICI and Oxford if some funding could be found to support these scientists. Lindemann was quite persuasive and the first of these ICI grants (for £400 per year) went to Mendelssohn starting on May 1, 1933.

Now that funding had been identified, Lindemann started work on bringing people to Oxford. At Lindemann's request, Mendelssohn corresponded with the rest of the Breslau group. It was assumed that the Nazis were reading the mail and in order to obscure what they were doing a rudimentary code was used. The letters between Oxford and Breslau took the form of enquiries regarding a series of air compressors that the Clarendon Laboratory was interested in buying. The high-pressure compressor referred to Simon while the low-pressure compressor referred to Kurti. The working pressure of the compressors in atmospheres stood for the annual salary in English pounds. In this way, negotiations were carried out between Oxford and Breslau via Mendelssohn. Simon was also looking to place Heinz London at Oxford, and London asked if he could be known as the "vacuum pump" in the letters.

In the fall of 1933, Simon, Kurti and London all immigrated to England and took positions at Oxford. Heinz London moved to the University of Bristol in 1936. The last member of the Breslau group, Rosteslav Kaichev returned to Bulgaria and had a successful career as a physicist there.

Kurt Mendelssohn's father was Jewish and also under threat from the Nazis. Fairly soon after arriving in England, Kurt Mendelssohn brought his parents over as well. At the start of the Second World War in 1939, his father was not a British subject and was interned as an enemy alien on the Isle of Man for a few months of the war. His father and mother spent the rest of their lives in England.

With his family safe and a position, however tenuous, at Oxford, Dr. Mendelssohn could start to put the events of 1933 behind him. It was time to get to work.

Chapter 4
All Things Super: Oxford 1933–1939

*"We forgot about superconductivity for a while over a series
of rather pretty experiments with the helium film"
"The Clarendon Laboratory", Cryogenics
K. Mendelssohn 1966.*

The pre-war years in Oxford were pivotal for Kurt Mendelssohn (Fig. 4.1). During this time, he started a family, became a British subject, put down roots in Oxford and established himself as a talented researcher. Some of his most important scientific work was carried out in this period.

His initial reaction was relief that he and his family were safe. Writing after the war, he says that on his first night in England he slept soundly for the first time in many weeks. While he may have been frustrated by the fact that Simon was the senior man, he and Simon quickly worked out an arrangement. Dr. Mendelssohn and his students would study superconductivity and later, superfluid helium. Dr. Simon, Dr. Kurti and their students would research other topics in cryogenics, particularly the recently developed technique of adiabatic demagnetization which allowed the study of phenomena below 1 K. This division of the cryogenic topics held up with only a few problems for the rest of their careers at Oxford.

When Simon came to Oxford, he was able to bring with him from Breslau quite a lot of his laboratory equipment, including two helium liquefiers. Some of this equipment had never ever been used. The influx of people and equipment allowed the cryogenics group in the Clarendon Laboratory to get started quickly.

Kurt Mendelssohn's first significant work at Oxford was in superconductivity. This is the phenomenon in which some materials lose all electrical resistance when cooled below a certain temperature. The temperature at which a material becomes a superconductor is known as the critical or transition temperature. Under these conditions, the material can carry electrical current without any voltage drop due to resistance. To put it another way, if an electrical current were started in a closed

---

The original version of this chapter was revised. The correction to this chapter is available at https://doi.org/10.1007/978-3-030-61199-6_11

© The Author(s), under exclusive license to Springer Nature Switzerland AG 2021,
Corrected Publication 2021
J. G. Weisend II, G. T. Meaden, *Going for Cold*, Springer Biographies,
https://doi.org/10.1007/978-3-030-61199-6_4

Fig. 4.1 Kurt Mendelssohn
in his Oxford Laboratory
circa 1933–1935. Note the
open cryostat in the center of
the figure

loop of superconducting material it would continue to circulate forever without the
need for additional energy. Such a phenomenon was of great fundamental and
practical interest. At the time, the only superconductors that had been discovered
required cooling to around liquid helium temperatures (4.2 K).[1]

Superconductivity had been discovered by Kamerlingh Onnes in 1911 yet physicists in the 1930s were still a long way from understanding the behavior of
superconductors.[2] Mendelssohn and his coworkers were able to conduct experiments that extended their understanding.

One thing that everyone did understand was that superconductors could lose their
superconductivity, even below their critical temperatures, if they were exposed to a
high enough magnetic field. This is known as the critical field. As previously
mentioned, while in Breslau, Kurt Mendelssohn became interested in the effect of
a magnetic field on superconducting tin. Two Dutch researchers (Keesom and van

---

[1]In 1986 "high temperature" superconductors were discovered which become superconducting at
around 100 K.

[2]The Bardeen–Cooper-Schrieffer theory which fully explains low temperature superconductivity
was not developed until 1956.

den Ende) had discovered a sudden change in the specific heat of tin when it lost its superconductivity. They also reported low values of specific heat just below the transition temperature of tin which they claimed was due to the presence of a magnetic field. Kurt Mendelssohn's knowledge of thermodynamics and specific heat measurements developed at Berlin led him to the following realization: If you thermally isolate a sample of superconducting tin and remove its superconductivity by subjecting it to an external magnetic field greater than its critical field, the sample should get colder. In 1934, Mendelssohn and a student, Judith Moore, demonstrated this effect, known as the magneto–caloric effect. The results were published in *Nature* on March 17th 1934. It was Kurt Mendelssohn's first publication as a University of Oxford physicist.

Mendelssohn and Moore found that the amount of cooling ranged from a 0.05 K temperature drop when the sample started at 3.3 K, and to a drop of 0.33 K with the sample starting at 2.5 K. They had hoped that the magneto-caloric effect would be useful for providing cooling at temperatures significantly below that of liquid helium. However, adiabatic demagnetization, which was first demonstrated by McDougall and Giauque at the University of California at Berkeley in 1933 and improved upon by Simon and Kurti, was clearly the superior method. In this technique, the spins (an intrinsic quantum property) of the electrons in a magnetic salt are aligned in a magnetic field. Similar to the magneto-caloric effect, when the field is removed, the sample cools. The difference is that the amount of cooling is much greater than in the magneto-caloric effect. In their first experiment, Simon and Kurti cooled their sample down to 0.1 K. Adiabatic demagnetization and the study of phenomena at the ultra low temperatures made possible by this technique would be the mainstay of Simon's and Kurti's research at Oxford.

In 1934, Kurt Mendelsohn also started work on another topic in superconductivity, one that was to be much more fruitful. The previous year, Walther Meissner and Robert Ochsenfeld working in Berlin had discovered that, contrary to all expectations, when a superconductor in a magnetic field is cooled below its transition temperature the superconductor will expel all the magnetic field from its interior. This result, known as the Meissner Effect, implied that materials in the superconducting state do not allow magnetic fields to penetrate them. Such a material is known as a perfect diamagnet.

Mendelssohn working with J. D. Babbitt, who was a Canadian Rhodes Scholar, set out to reproduce Meissner's experiment. What they found was complicated but illuminating. They began with two tin spheres of the same size: one solid and the other 50% hollow. The spheres were cooled to below their transition temperature while held in a constant magnetic field. The solid sphere did in fact expel most but not all of its magnetic field while the hollow sphere expelled about half as much of the magnetic field as the solid one. It seemed that tin was not a perfect diamagnet but instead allowed some of the magnetic field to penetrate it.

In another set of experiments, the spheres were held at a temperature below their transition temperature and a magnetic field was increased external to the sphere. Mendelssohn and his coworkers found that the magnetic field did not penetrate the sphere until the field reached a certain value, which Kurt Mendelssohn termed the "penetration field". At this point, the field started to penetrate the sample. The amount of the field penetration increased until the critical magnetic field was reached at which point all superconductivity vanished.

The group carried out another set of experiments with tin and lead alloys and found that most of the magnetic field remained inside the alloys after they were cooled below their transition temperature and became superconducting. They found that for pure mercury all the field was expelled and thus superconducting mercury was in fact a perfect diamagnet.

Dr. Mendelssohn and colleagues at Oxford explained these results by theorizing that each sample contained intermixed regions of superconducting and normally conducting material and that the magnetic field penetrated into these normal regions and was trapped within them. This explanation, known as the "sponge model" could be shown to explain both the difference between the solid and hollow spheres as well as the results of the alloy measurements.

At the same time as these experiments were being carried out, Heinz London from Breslau and his older brother Fitz arrived as refugees in Oxford. The London brothers immediately started to develop a general theoretical model of superconductivity. There was a great deal of communication and cross fertilization between the experiments and theories of Mendelssohn's team and the theoretical work of the London brothers. This synergy benefited everyone.

Kurt Mendelssohn next investigated two thermodynamic properties (specific heat and entropy, which can be derived from the specific heat) of superconductors. The experiments made good use of the skills he had developed in Berlin measuring specific heats at cryogenic temperatures. Working with his student John Daunt, Mendelssohn first measured the variation of the critical magnetic field with temperature for lead, tin, mercury, tantalum and niobium superconductors. From this data, they were able to calculate the difference in entropy and specific heat between the normal and superconducting states of each material. The entropy difference was higher than predicted by current theory. Daunt and Mendelssohn hypothesized that the electrons in the superconducting state are in a more ordered (that is a lower entropy) arrangement than those in the normal state.

They next performed a clever experiment to directly test this hypothesis. Mendelssohn and Daunt constructed a small closed loop of lead. They fixed opposite sides of this loop to liquid helium baths operating at slightly different temperatures. A current was then induced in the lead. As lead is superconducting at these temperatures, the current will flow without any resistive losses. Mendelssohn and Daunt next reversed the current flow and checked to see if the temperature difference across the ring changed. They found that it had not and it was well understood that this could only be true if the entropy of the electrons carrying the superconducting current was zero.[3] This confirmed the results of the previous experiment. The full paper describing this experiment was not published until after the war. In it, Mendelssohn points out that the results also imply the existence of a gap in the energy between the superconducting and normal electrons.

---

[3]More precisely, they had found that the Thomson heat of superconductors was zero.

As it turned out, Kurt Mendelssohn's experiments were accurate and his explanations were not far wrong. We know today that low temperature superconductors may be divided into two types. Type I superconductors expel essentially all[4] magnetic field while superconducting. Type II superconductors have two critical magnetic fields: a lower critical field denoted H_{c1} and an upper critical field denoted H_{c2}. While in a magnetic field below H_{c1}, the superconductor acts just like a Type I and expels all magnetic field. The lower critical field is thus the same parameter that Mendelssohn called the "penetration field". Above the lower critical field but below the upper critical field, the superconductor remains superconducting but does allow magnetic field to penetrate through it. In this operating region, there are both normally conducting and superconducting parts of the material as explained by Mendelssohn's sponge model. However, it turns out that the normally conducting parts are microscopic in size and the correct explanation for this so-called mixed state of the superconductor was not provided until 1957 by the Russian physicist A. A. Abrikosov. Above the upper critical field, the entire material becomes normally conducting. Consistent with Mendelssohn's observations, Type I superconductors tend to be pure elements while Type II superconductors tend to be alloys. However, it should be pointed out that Mendelssohn's sponge model required that there be alloys or impurities in the material, whereas the correct explanation applies for completely pure materials such as niobium that are also Type II superconductors.

We also know today that when a material is superconducting some electrons are in the superconducting state and carry all the current while others remain in the normal state. As the temperature decreases, more and more electrons move into the superconducting state. Just as Kurt Mendelssohn predicted, the superconducting electrons are more organized (they are in fact paired up) than the normal electrons and there is an energy gap between the normal and superconducting states.

Mendelssohn's group was not alone working on these problems. Superconductivity was a dynamic research topic at the time and groups at Leiden in the Netherlands, Moscow and Kharkov in the USSR and Cambridge (who had gotten their helium liquefier into operation) among others, all made vital contributions. Nevertheless, Mendelssohn's work was substantial and helped establish his reputation.

All his experiments were carried out using the Simon expansion method for helium liquefaction. This required that each experimental apparatus contain its own Simon liquefier. As Mendelssohn had learned in Berlin, such a requirement provided excellent hands-on training for the students.

The Simon liquefiers required cooling with liquid hydrogen. The experiments were generally set up and cooled down during the day with the actual helium liquefaction and data-taking occurring late at night. As a result, Mendelssohn developed the habit of working during the day, going home for dinner and returning to the laboratory (frequently bearing sandwiches prepared by Jutta for the students)

---

[4]Type I superconductors do allow magnetic field into a microscopically thin layer on the surface. This is known as the penetration depth.

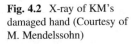

Fig. 4.2 X-ray of KM's
damaged hand (Courtesy of
M. Mendelssohn)

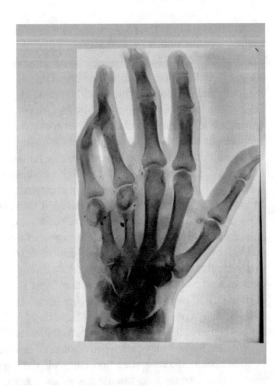

to perform the experiment. He would then work until 2 or 3 in the morning and sleep
late the next day having his breakfast in bed. This was a research schedule that he
maintained for many years. His children remember him playing records of Wagne-
rian opera upon his return from the laboratory in the middle of the night and of being
instructed not to wake their father up in the morning.

During preparations for one of these experiments Kurt Mendelssohn was seriously injured while trying to pressure test a small steel vessel which he held in his
hand and was unlucky enough to demonstrate its faultiness by an explosion which
badly damaged his right hand (Fig. 4.2) and removed him from the laboratory for
several months. The surgeon's repair work was imperfect, though subsequently he
managed remarkably well in spite of two crippled fingers.

Throughout this period in the 1930s, Kurt Mendelssohn's position at Oxford
University was uncertain. He was supported by a series of grants from the Imperial
Chemical Industries that were renewed every few years, so he was continually on the
lookout for a permanent position at other institutions. However, there was now a
flood of academic refugees coming to England from the continent. In response to
this, English academics started the Academic Assistance Council to help place the
refugees at least in temporary positions at British institutions. Over the course of the
Second World War, the council helped 2600 refugees from European universities.
These numbers meant that there were far more qualified people than permanent
positions and as a result, KM remained at Oxford in a temporary position. Ultimately, some of the positions he applied for he was better off not having. For

Fig. 4.3 A painting of 235 Iffley Road, Oxford (Courtesy of M. Mendelssohn)

example, he applied for a position as a physics professor at Raffles University in Singapore. The Japanese at the beginning of the Second World War, of course, invaded and occupied Singapore.

Despite his temporary status at Oxford University, Dr. Mendelssohn was clearly happy with his work and life in England. He had a very fine house (Fig. 4.3). As daughter Monica said "Kurt and Jutta moved into a large house on Iffley Road, just before the arrival of their first child, Corinna in December 1937. From then on the family expanded rapidly, Ursula February 1940, Monica January 1943, Diana May 1945 and finally James October 1946. They also had had a son David born in August 1941 who sadly died in November 1941. Kurt was proud of his large family, but unfortunately all the photos of Jutta and the children were taken by him so there are no pictures of him with his family."

Fig. 4.4 Jutta Mendelssohn and the Mendelssohn Children—1947 (Courtesy of M. Mendelssohn)

The photograph in Fig. 4.4 is an example, taken in 1947. Monica Mendelssohn writes that "it shows all the children ready, at about 6.30 a.m., to set off on the long drive to Cromer in Norfolk for their summer holiday. As a child Kurt had spent his holidays on Nordeney, one of the Baltic islands, and his fondness for the island was re-kindled by the sand and the dunes of the Norfolk coast."

Dr. Mendelssohn would keep this house for the rest of his life and it would be the focal point of family activities. Iffley Road was fondly remembered by many of his students who were warmly entertained there for group parties hosted by Kurt and Jutta and it was also an occasion to meet the children. He and Jutta would frequently welcome colleagues and foreign visitors at the house.

Kurt Mendelssohn became a naturalized British subject in 1938. As a consequence, when war with Germany started in 1939, he could continue to work in Oxford instead of being sent to an internment camp.

Another "super" phenomenon was at the heart of Mendelssohn's other area of research in the Thirties. By 1930, it was known that as the temperature of liquid helium was reduced below 2.2 K there was a sudden spike in the specific heat of the liquid indicating that there was a transition to another phase of liquid.[5] By convention this second phase of liquid helium was labeled helium II (He II) while the liquid phase above 2.2 K was denoted helium I (He I). The spike in specific heat as a

---

[5]The most well-known phases of materials are: solid, liquid and gas. Ice melting into liquid water is a change in phase as is water boiling into steam. Nature is more complex than this however; for example depending on the pressure and temperature there are a number of different phases of solid ice.

function of temperature resembles the Greek letter lambda and thus 2.2 K became known as the lambda point. While helium II was known to exist, not much was known about its properties until the mid-thirties. In 1937, J. F. Allen of Cambridge reported that the heat transferred through helium II was vastly higher than that in helium I, and that in He II the heat transfer was not linearly dependent on temperature as would be expected. Shortly thereafter, Allen and P. Kapitsa (who was now in Moscow) independently discovered that the viscosity (which can be thought of as a fluid's resistance to flow) of He II was extremely small and perhaps even zero. This property is known as superfluidity and He II also became known as superfluid helium. The property of superfluidity occurs only in liquid helium.

Mendelssohn's study of superfluid helium started just after Allen's first paper was published and like many good things in science came about accidentally. Mendelssohn and his student John Daunt were studying the effect of infrared light on superconductors. This required that they use a glass dewar to allow the light to shine on the superconductor. As a side benefit, it also allowed them to see the liquid helium cooling the superconductor. They started to make visual observations of the helium. They noticed that once the helium was below 2.2 K, a thin film of helium would climb up the wall of the dewar. Here was fluid moving upwards against the force of gravity.

Mendelssohn and Daunt quickly realized that studying the behavior of this film was far more interesting than their current superconductor work. The ability to recognize that a new observation is worthy of further study and to change research directions accordingly is a sign of a talented scientist.

The work of Dr. Mendelssohn and his students on helium II films[6] is an illustrative example of Mendelssohn's meticulous contributions to cryogenic research. The work is some of Mendelssohn's most inventive and fundamental and addressed a very recently discovered (in part at least by Mendelssohn) phenomenon. This work was a significant justification for Mendelssohn's later scientific awards. The film experiments started in the late 1930s and were continued after World War II (see Chap. 6).

The helium II film research also provides an early example of Mendelssohn's ability to present complicated subjects in a very clear straightforward manner making good use of both graphical representations and textual summaries. Thus, the He II film papers written by Mendelssohn can be seen as stylistic precursors to his later books (Chap. 9).

There were earlier indications of the film phenomenon. H. Kamerlingh Onnes of the University of Leiden, who was the first person to liquefy helium, reported in 1922 that the levels in concentric dewars containing liquid helium would quickly become equal [1]. In 1936, Simon and his student B. Rollin had noticed that the heat leak to a vessel containing He II was much higher than expected [2]. They theorized

---

[6]In their early papers, Mendelssohn and his team referred to this phenomenon as the "transfer effect". Later, as more understanding was gained, the research community including Mendelssohn adopted the term "film effect".

that this was due to heat transfer through a thin film of helium that covered the wall of the tube connecting the vessel to the region outside the vessel. Kurt Mendelssohn himself had observed anomalous results during thermal measurements below 2 K which seemed to be related to helium deposited on the surface of his experiment.

The helium film and its behavior had now been directly observed in Kurt Mendelssohn's experiment on infrared light and superconductors. He and his students set up a series of elegant experiments to study helium films.

The concept of elegance in the design of an experiment is probably unfamiliar to nonscientists. An elegant experiment is one which provides an accurate, reproducible answer to a well posed scientific question. The experiment should be designed to provide unambiguous results within a known level of accuracy. In investigating a totally new phenomenon, such as helium films, it is generally best to design experiments to answer one question at a time and then introduce additional features or build new experiments to answer subsequent questions. These questions may change over time depending on the results of earlier experiments. This is the approach that Mendelssohn and his team took. Notice also, that properly defining the question to be asked can be as important as the experiment itself. Designing an experiment to answer "what is the behavior of helium films?" could lead to a very complicated apparatus that might mask important results or produce spurious results that are hard to interpret. Asking the more focused question, "what is the change of levels between two concentric vessels containing helium below 2 K?" results in a more elegant experiment and again this is the approach that Mendelssohn and Daunt took.

The apparatus they designed is shown in Fig. 4.5. The innermost bath (V) contains the He II and is the location of the experiments. The complexity of the apparatus is driven by the need to reduce as much as possible the heat leaking into bath V from the outside world. Thermal (or infrared) radiation is almost entirely absorbed by two liquid baths; one with liquid nitrogen operating at 77 K and one containing liquid hydrogen at 20 K. All the solid connections between room temperature and volume V first pass through a liquid helium bath at 4.2 K which absorbs most of the thermal conduction heat prior to volume V. Great care was taken from the beginning to minimize the heat leaks from the outside world as it was recognized that such heat leaks would evaporate the helium film under study and affect the results. The bottom half of the apparatus was constructed from glass to allow visual measurements of the film.

The first experiment carried out involved two interconnected vessels both containing He II as shown in Fig. 4.5. Over a short period of time the level in the upper vessel drops and the level in the lower vessel rises. However, in this first experiment much more helium leaves the upper vessel than goes into the lower vessel. This is determined to be due to much of the helium leaving the upper vessel by evaporation and transport as vapor to other higher parts of the apparatus. A question that remains is "does the helium that moves to the lower vessel do so due to the newly observed film effect or through He vapor; that is evaporation in the upper vessel and condensation in the lower vessel?" This is tested by adding a wick that connects the upper vessel through the connecting tube to the lower vessel. The

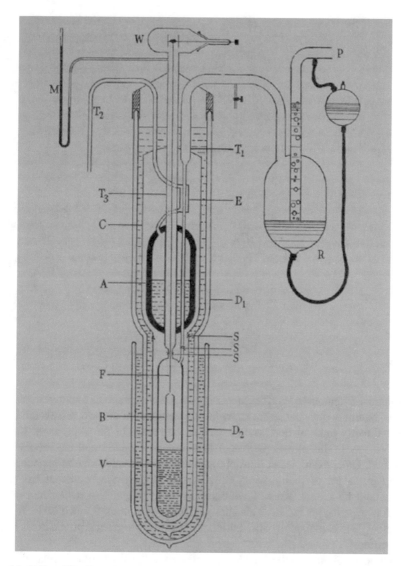

Fig. 4.5 Helium Film Experimental Apparatus [3] V—Experimental Volume, D_1—Liquid Hydrogen Bath, D_2—Liquid Nitrogen Bath, A—Liquid Helium Bath

results here are very clear. With the wick in contact with the liquid helium in the upper vessel, the rate of transfer of helium between the upper and lower increases. Once the liquid level in the upper vessel drops below contact with the wick, the rate of helium transfer also drops. From these simple experiments, it can be shown the rate of helium transfer is related to the width of the surface connecting the two vessels. Thus, the helium transfer between the vessels is done by the He II film flow on the solid surface connecting them not via evaporation.

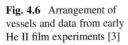

Fig. 4.6 Arrangement of vessels and data from early He II film experiments [3]

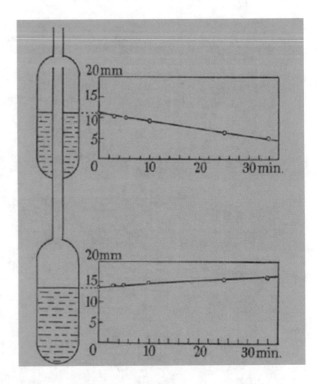

Figure 4.6 demonstrates Kurt Mendelssohn's use of graphics to convey scientific results. Not only does the figure show the arrangement of the two vessels but it also shows directly links of that arrangement to the results of the experiment (helium level vs. time). This allows the reader to more easily understand the results being presented. Other examples of this approach will be seen below and in his books. He clearly put a lot of thought into how best to display the results of his work. Throughout his career, Kurt Mendelssohn would use graphics rather than mathematics to describe his results. This approach, of course, made it much easier for him to present clear explanations of scientific principles to the general public in his books and writing.

The observation that the transfer of helium via the He II film is related to the surface connecting the two vessels implies that larger, more easily observable effects, can be created if the surface connecting the vessels is made larger. This led to the next experiment. The smaller container is suspended on a cable that allowed it to be lifted in and out of the larger volume (V) in the apparatus (see Fig. 4.7).

The experiment was started by partially filling the outer container (V) with He II. Mendelssohn and Daunt discovered that when they lowered the empty inner container partially into the helium in the outer container, a film of He II would creep up the outer wall of the inner container and fill it until the levels in both containers were equal. If the inner container was now lifted further out of the larger one, the helium

Fig. 4.7 Mendelssohn's
and Daunt's Movable
Vessel Experiment in He II
Film Flow [4]

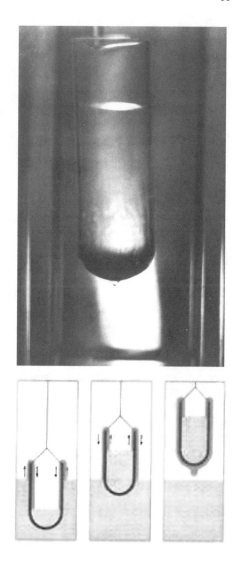

would flow up the wall of the inner container and out into the outer container until the levels were again equalized. If the partially filled inner container was completely pulled out of the helium in the outer container, the helium in the inner container would flow up the walls of the container and drip into the outer container until all the helium in the inner container was gone. See Fig. 4.7.

These experiments were continued by Mendelssohn and Daunt throughout the rest of the 1930s investigating all aspects of He II film flow including measuring the thickness of the film, the effect of level difference on the transfer rate and the relationship of the film transfer to heat transfer between the vessels. Figure 4.8 shows the results of an experiment investigating the effect of level differences on the

Fig. 4.8 Observed
Variation of Film Transfer
Rate with Level Differences
[3]

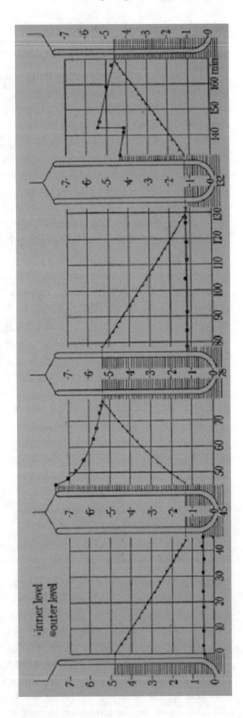

rate of transfer through the film. Note that as in the case of Fig. 4.5, Mendelssohn combines experimental data with graphics to describe the results.

In 1938, Mendelssohn and Daunt published two consecutive papers [3, 5] in the *Proceedings of the Royal Society* that summarized their work to date. These papers not only included the Figs. 4.5 and 4.7 above but also used an approach in which what had been learned from the series of experiments was summarized by a set of short declarative sentences at the end of each paper. This technique makes it much simpler for the reader to understand what has been discovered. It is worth quoting these sentences in their entirety to show what Mendelssohn and Daunt had found as of 1938.

1. "Liquid He II collects always at the lowest available level."
2. "This transfer takes place in a film along the solid surface of the containers"
3. "The rate of transfer is practically independent of the difference in the height between the levels and of the underlying material"
4. "The rate of transfer is exactly proportional to the width of the connecting surface"
5. "The transfer from one container to another is restricted by the narrowest part of the connecting surface above the height of the upper level. At places below the upper level, liquid can collect in drops from the film."
6. "The heat conductivity of the transfer film is small. The high heat transport in the vessels containing liquid He II is due to the transfer of helium along the walls."
7. "The thickness of the transfer film is about 3.5×10^{-6} cm."
8. "The rate of transfer changes with temperature. The change has been determined between 1 and 2.2 K." [3, 5]

These are clear, concise sentences. Even someone unfamiliar with cryogenics and He II could understand what had been discovered from this summary.

Other groups, of course, were studying the He II film behavior and Mendelssohn returned to this work after World War II.

At about the same time that Kurt Mendelssohn was conducting his first film experiments, J. F. Allen in Cambridge built a small container with a heater inside and two openings, one blocked by fine emery powder and the other open. He found that if they immersed the container in a bath of He II so that the open end was above the liquid and the blocked end was in the liquid, when the heater was turned on He II would flow through the tiny openings of the powder and spurt out through the other opening. This unexpected result was known as the fountain effect or thermo-mechanical effect. Mendelssohn and Daunt then observed the same phenomena in their film tests. They put a heater in their inner container and lowered the container into a He II bath. Once the levels in both the inner and outer container were equal, they turned on the heater and found that helium flowed through the film into the inner container and raised its level above that of the outer container.

A theory was being developed to explain the odd behavior of He II.

L. Tisza proposed the two-fluid model which treats He II as consisting of two interpenetrating fluids: one (the normal fluid component) had finite viscosity and finite entropy and the other (the superfluid component) had zero viscosity and zero entropy. The interaction of the two components could be used to explain the behavior of He II. Fritz London thought that the superfluid component came about due some of the fluid dropping down into the lowest quantum energy state. Heinz London was able to build on these models to explain the thermodynamics behind the fountain effect.

Heinz London also realized that the fountain effect should, in effect, be able to run backwards.[7] Essentially, in the fountain effect, raising the temperature of the He II in the small container caused helium to enter the container through holes in the emery powder. If conversely one removed He II from the container, the temperature in the container should rise without any additional heating. Allen, Daunt and Mendelssohn immediately started searching for this phenomenon. Daunt and Mendelssohn won this race using their film-flow apparatus. They replaced their inner container with one having a small hole blocked with emery powder and immersed it in a He II bath. When they pulled the container out of the bath, helium would run out of the holes in the powder plug and the temperature inside the inner container would rise. This was called the mechano-caloric effect of He II and Mendelssohn and Daunt are credited with its discovery.

The unique behavior of He II was discovered and explained very quickly. Kurt Mendelssohn described the atmosphere of the time in his later book *The Quest for Absolute Zero*: "The attack on the problem is an excellent example of scientific co-operation at its best. There was fierce competition between the different laboratories but they kept each other fully informed of the progress achieved, often over the telephone and through personal meetings."

In many ways, the model for the behavior of He II is similar to that of the behavior of the electrons in a superconductor. Writing in 1942, Mendelssohn and Daunt would point out these similarities. In 1969, Mendelssohn would describe both superconductivity and superfluidity as "condensations in momentum space". While this formularization was never really accepted, it does illustrate Mendelssohn's tendency to try to fit his results into a broader physical picture.

Dr. Mendelssohn's research output during the thirties is impressive. He published 32 research papers in highly respected journals such as *Nature* and the *Proceedings of the Royal Society*. He made fundamental discoveries in the areas of superconductivity and superfluid helium. Much of his later honors, such as being made a Fellow of the Royal Society, were based on this work. His work in this period also shows tendencies that are repeated throughout his career. Mendelssohn was capable of designing elegant experiments that can fully answer the question at hand. This is illustrated by the determination of the zero entropy of electrons in superconducting currents and the He II film experiments. He was able to recognize and quickly pursue fruitful areas of research. His abandonment of the study of infrared absorption by superconductors for the far more fundamental studies of He II films is telling. Another example, was his scooping of J. Allen and the Cambridge group by

---

[7]That is, it should be thermodynamically reversible

Fig. 4.9 Kurt Mendelssohn in front of the Old Clarendon Laboratory, Oxford (Courtesy of J. Mendelssohn)

discovering the mechano-caloric effect in He II. In the laboratory, like all good scientists, Mendelssohn was an opportunist.

His talent in describing his results was also evident. His papers were clear and direct. He used innovative graphics to clearly illustrate the results of the He film experiments. This ability echoes his skill as a lecturer and would serve him well when he started to write books intended for the general public.

While his excursions into theory were not always successful and his models, like the sponge model, were not always correct, they were frequently not very wrong. More importantly, his experimental results and models stimulated work by gifted theoretical physicists such as the London brothers.

Kurt Mendelssohn kept current with the work going on in physics and was on personal personnel terms with experts around the world. His capacity to know who was doing what and his recognition that such relationships were important would play a major role in both his scientific career and the rest of his life.

In late 1939, the Clarendon Laboratory started to move from the old Venetian Gothic construction (Fig. 4.9) that had housed it (to a new larger building adjacent. The fact that larger quarters were required is a testament to Professor Frederick Lindemann's success in building up physics research at Oxford. However, all research was stopped during the move, and with the beginning of World War II in 1940 all work in the Clarendon was devoted to war research. Kurt Mendelssohn would not go back to cryogenic research until after the war ended.

References

1. H. Kamerlingh-Onnes, Comm. Phys. Lab. Univ. Leiden, No. 159 (1922)
2. B.V. Rollin, Act. 7 Int. Congr. Refrig. No. 1 (1936)
3. J.G. Daunt and K. Mendelssohn, "The Transfer Effect in Liquid He II 1. The Transfer Phenomena", Proc. Roy. Soc. A. 179, 423, (1939)
4. K. Mendelssohn, *The Quest for Absolute Zero*, World University Library (1966)
5. J.G. Daunt and K. Mendelssohn "The Transfer Effect in Liquid He II, II. Properties of the Transfer Film." Proc. Roy. Soc. A. 170, 439, (1939).

Chapter 5
War and Controversy: 1940–1945

*"I must definitely refuse to be either bought or bullied into a
course of action of which I do not wholly approve..."*
K. Mendelssohn—1942 [1]

Like Winston Churchill, F. A. Lindemann had an American mother, conservative politics, inherited wealth, and an early appreciation for the threat that Nazi Germany posed to England. It is not surprising then that Churchill and Lindemann would work closely together throughout the war. In 1939, Lindemann, while retaining his position as Head of the Clarendon Laboratory, started work in the Admiralty as Winston Churchill's personal assistant and head of the Statistical Section. When Churchill became Prime Minister in 1940, Lindemann remained his assistant and later became the paymaster-general and a privy counselor. However, his most important function was as a scientific advisor to Churchill and one of his inner circle. As a result, the Clarendon Laboratory had very high connections in the government and during the Second World War, Frederick Lindemann devoted all the facilities and talents of the laboratory to military work. Laboratory research was centered on developing RADAR systems (radio detection and ranging) for anti-submarine and anti-air warfare. Cryogenic research was over for the duration of the war.

At first, the refugee scientists from Germany were not trusted with war work. Later, Francis (formerly Franz) Simon and Nicholas (formerly Nicolaus) Kurti became heavily involved in the British contributions to the development of the atomic bomb (whose British code name was the Tube Alloys Project). Simon and Kurti developed important techniques for separating out the uranium isotope ^{235}U required for the bomb via gaseous diffusion. Their efforts were successful and gaseous diffusion became a standard technique for the separation (or enrichment) of uranium. Simon was knighted as Sir Francis after the war largely for his contributions to this project.

The original version of this chapter was revised. The correction to this chapter is available at
https://doi.org/10.1007/978-3-030-61199-6_11

J. G. Weisend II, G. T. Meaden, *Going for Cold*, Springer Biographies,
https://doi.org/10.1007/978-3-030-61199-6_5

Fig. 5.1 Schematic of
Oxford Vaporizer [2]

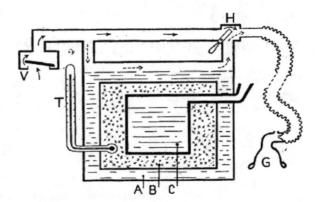

Kurt Mendelssohn became interested in medical physics. In early 1940, Prof. R. R.
Macintosh of the University of Oxford Nuffield Department of Anesthetics
approached Kurt Mendelssohn seeking help with a problem in anesthesiology.

The most common anesthetic of the time was ether vapor which was mixed with
air by the anesthesiologist to provide the correct level of sedation. The problem with
this technique was that unless the ether liquid was kept at a constant temperature the
rate at which the ether vapor was produced varied. In hospitals, up to that time, the
standard approach was to wrap something warm around the bottle of ether. The rate
of production of ether vapor would still vary but the anesthesiologist would use his
experience to adjust the ether/air mixture to keep the amount of sedation constant. In
wartime, with the need to provide anesthesia in large numbers of field hospitals and
forward areas which would not all have trained anesthesiologists, this solution would
not work. A simple machine that could be safely operated by people with only
limited training was required. This was the problem that Prof. Macintosh brought to
Mendelssohn.

The key to the solution was keeping the ether at a constant temperature so that the
vapor was evolved at a constant rate. To someone with Mendelssohn's training in
thermodynamics, the solution was obvious. Materials stay at a constant temperature
when they are undergoing a first order phase transition. Examples of such transitions
include boiling, melting or freezing of water. A first-order phase transition is one that
requires energy to proceed, which results in the material staying at a constant
temperature during the course of the transition. For example, no matter how much
heat you put into a pot of boiling water at atmospheric pressure it will stay at a
temperature of 100 °C as long as it is boiling. Any extra heat goes to speeding up the
boiling, not raising the temperature. To solve the problem, Mendelssohn had to find
a first-order phase transition that would give him a constant temperature in the range
needed to vaporize ether.

The solution was to use the melting of hydrated calcium chloride crystals. These
crystals melt at 30 °C (303 K) and will stay at that temperature as long as the melting
goes on. They provide the constant temperature required for the vaporization of the
ether. A schematic of the ether vaporizer is shown in Fig. 5.1. The ether bath
(A) surrounds a sealed container of the calcium chloride crystals (B). This container,
in turn, surrounds a reservoir into which boiling water is poured. The heat from the

water is transferred to the crystals causing them to melt. As long as the crystals are melting, their temperature and the temperature of the surrounding ether remain at 30 °C. The constant temperature ether in turn provides a constant concentration of ether gas which may be mixed with air via valve (H) and sent to the patient through the mask (G).

The advantage of this design is that the melting crystals act as a temperature reservoir. As the heat is transferred from the boiling water to the crystals, the water will stop boiling and become cooler.

However, as long as the water temperature remains above 30 °C, the crystals will continue to melt and the ether bath will remain at a constant temperature. Once the water temperature drops below 30 °C, the melting will stop and because heat is still being transferred from the crystals to the ether, the temperature of the crystals will drop. The temperature drop will be shown on the thermometer (T) and will indicate to the operator that the hot water reservoir must be refilled.

The resulting device was beautifully simple. It was portable. It required no power, only boiling water for operation. The calcium chloride crystals may be melted and refrozen again and again. The only instrumentation built into the device is a thermometer for the crystal temperature and a gauge to indicate the amount of ether remaining. With the addition of a calibrated valve to control air/ether concentration, medics with limited training in anesthesiology could safely operate the vaporizer.

While the theory behind the device was elegant, a substantial amount of detailed engineering design was required. The amount of calcium chloride crystals needed, the size of the water and ether reservoirs and the connecting piping all had to be determined. Ensuring proper heat transfer between the water, crystals and ether within as compact a package as possible was particularly critical. Kurt Mendelssohn's experience in designing specialized experimental equipment involving fluids and heat transfer made him perfectly suited for this work. He applied techniques such as using materials with good thermal conductivity and creating surfaces with fins on them to improve heat transfer. It is by no means a coincidence that the compact bath within a bath design finally used for the vaporizer resembles the standard design of cryogenic dewars. Figure 5.2 shows the resulting Oxford vaporizer.

A second, somewhat more sophisticated, version of the Oxford vaporizer was also produced [3]. However, this device which operated at a higher pressure, was better suited to situations in which there was a trained anaesthetist [4].

A patent application was made for the vaporizer in August of 1940 and granted in December of 1941. The Oxford vaporizers proved to be extremely successful with thousands produced by the Morris Motors company of Oxford. They continued to be used by the British military long after the end of World War II.

However, in 1941 and 1942, a dispute developed between Mendelssohn and Macintosh regarding the vaporizer. The key issue was who had made the original suggestion about the use of the latent heat of crystals to provide a constant temperature and subsequent complications concerning the patent taken out by Nuffield. The matter was investigated but came to no particular conclusion in the end.

Fig. 5.2 The Oxford
Vaporizer. Note the compact
design and simplicity of the
control (courtesy of the
Wood Library-Museum of
Anesthesiology,
Schaumburg, Ilinois)

This dispute was unusually hostile and a more typical description of Mendelssohn's challenging side is described by David Shoenberg in his Royal Society Biographical Memoir

> Over such questions of priority and credit Mendelssohn often displayed undue sensitivity . . . he felt he had to be one-up and if, as often happened, he was in the right, his tendency to rub it in was not popular with his opponents. When, however, he was wrong he hated to admit it and tended to shift his ground. In a sense perhaps it was this one-upmanship which was the spur to his scientific work and was responsible for the ingenuity of his experiments. He felt he had to be one-up on Nature and often told his students that the whole art of research is to outwit Nature. Although some of the disputes in which Mendelssohn was involved were sometimes acrimonious, he was basically warm hearted and often it was not difficult to heal a temporary breach and restore cordial relationships. [5]

Part of the tension in this dispute likely stems from the fact that Kurt Mendelssohn, a physicist, was working in medical research. Physicists had only recently begun to involve themselves with medical problems with the discovery of X-rays. Today, there are many programs in Medical Physics or Biomedical Engineering, but in the 1940s the concept that laws of physics (such as thermodynamics) could profitably help solve problems in medicine was still quite new.

Another example of Dr. Mendelssohn's application of physics to medical problems is seen in his and D. S. Evans's work on the measurement of infrared radiation. Radiant heat was used for a variety of medical therapies, but at the time, the amount of heat provided to the patient was not necessarily well understood or measured. In fact, Mendelssohn and Evans were able to show that in at least one system for radiant therapy, the heat reflectors did not do a very good job of reflecting heat but instead heated up themselves and became secondary heat sources that produced more radiant heat than the primary heaters [6]. In order to address this problem, Mendelssohn and

Fig. 5.3 Kurt Mendelssohn
in the Home Guard
(Courtesy of
M. Mendelssohn)

Evans developed an instrument that could correctly measure the heat being applied to a patient [7].

Recall that Kurt Mendelssohn had earlier been examining the effect of infrared radiation on superconductors when he switched his attention to studies of the He II film (Chap. 4). Thus, he had already been thinking of ways to accurately measure incident infrared radiation.

During the rest of the war, Mendelssohn continued to work in the area of medical physics. He and his various collaborators studied a wide range of topics including: subcutaneous skin temperature under burns, the accurate measurement of pressure in sinuses, and the transmission of infection during blood withdrawal. Mendelssohn's work on the physics of blood pressure measurement led him to develop a prototype of a new blood pressure measurement system which was tested clinically. The common thread running through all these projects is the application of physics principles or measuring techniques to medical problems.

Mendelssohn's work during the war was the first example of his applying his expertise in physics and engineering to problems outside his immediate area of study. Later in life, he would do this in areas as diverse as Egyptology and science policy. Thus, Mendelssohn's experience during the war years not only illustrates his concern that he always be given proper credit for his work but also his interest in interdisciplinary research.

Throughout the war, Kurt Mendelssohn served in the local detachment of the Home Guard (Fig. 5.3).

The following anecdotes were kindly provided by Monica Mendelssohn: "Besides taking up research in medical physics KM was conscripted into the Home Guard. This was a very strange combination of carrying out high precision science in one mode and extremely basic army training in the other. KM was not natural army material and as he pointed out the dictionary definition divided intelligence into three categories, human, animal and military.

His army experience included him in going on exercises of one type or another and he gleefully remembered the 'Gunnery Practice'. This took place somewhere in Wales where a statuary group of smart looking officers were seated to observe the quality of the activity. All was set up and it was KM's duty to call out the instructions for firing and over the loudspeaker system came the precision commands 'Ready', 'Aim', 'Fire' in an undeniably thick guttural German accent."

References

1. K. Mendelssohn, Personal Papers, Special Collections & Western Manuscripts, Bodleian Library, University of Oxford.
2. H.G. Epstein, R.R. MacIntosh and K. Mendelssohn, "The Oxford Vaporiser No. 1", The Lancet, July 19, 1941.
3. S.L. Cowan, R.D. Scott, S.F. Suffolk, "The Oxford Vaporizer No 2.", The Lancet, July 19, 1941.
4. H.G. Epstein, E.A. Pask, "The Performances of the Oxford Vaporisers with Ether", The Lancet, July 19, 1941.
5. D. Shoenberg, "Kurt Alfred Georg Mendelssohn", Biographical Memoirs of the Fellows of the Royal Society, Vol. 29 (1983).
6. D.S. Evans and K. Mendelssohn, "The Physical Basis of Radiant Heat Therapy", Proc. Roy. Soc. Med. 38 (1945).
7. D.S. Evans and K. Mendelssohn, "The Measurement of Infra-Red Radiation for Medical Purposes", J. Sci. Instrum. 23, 94 (1946)

Chapter 6
Studies and Students Oxford 1945–1973

"A significant feature of the post-war years was the influence that the Oxford low temperature school began to have on research in other laboratories. A great number of its former research students set up low temperature work in various parts of the world, creating schools of their own"
"The Clarendon Laboratory, Oxford, (The World of Cryogenics. IV.)" K. Mendelssohn, Cryogenics Volume 6, No. 3 (1966)

Some of the most interesting aspects of Dr. Mendelssohn's professional life involved ancillary activities such as travel, writing and establishing mechanisms for communications within the broader cryogenics community. These are covered in Chaps. 7–10. These activities, however interesting, should not obscure a basic feature of Kurt Mendelssohn's life. For 40 years, he was a researcher and lecturer in cryogenics at Oxford University. During this time, he published almost 200 papers and articles and supervised 50 graduate students. Appendix 1 provides lists of his publications and students.

Just as before World War II, Dr. Mendelssohn's cryogenics group was part of the Clarendon Laboratory (Fig. 6.1). Keep in mind that there was also another, very active, cryogenics group, led by Francis Simon and Nicholas Kurti in the Clarendon Laboratory. This group stressed research on nuclear demagnetization experiments that enabled the reaching of extremely low temperatures down to the micro-Kelvin regime. The existence of these two strong research groups helped make Oxford one of the leading world centers for cryogenic research during the remainder of the twentieth century.

In 1949 came invitations to four of the country's primary research workers in low temperature physics to give a series of lectures at the Royal Institution in the City of Westminster. This worthy organization, with its Michael Faraday Museum, in the heart of Central London had been founded for the purpose of providing scientific education via demonstrations and lecturing at a level to appeal to the inquiring masses about the latest advances in science. The lectures took place in February and March 1950 and were given by four of the leading figures in low temperature physics research including Kurt Mendelssohn who lectured on superconductivity. The others

Fig. 6.1 The Clarendon Laboratory in 1949 (Courtesy of B. S. Chandrasekhar)

were Francis Simon, F.R.S. who surveyed the current problems in low temperature physics, Nicholas Kurti about research at temperatures lower than one degree absolute, and J.F. Allen, F.R.S. on liquid helium. The subsequent book [1] of the four lectures was published by Pergamon Press 2 years later. It made a lasting authoritative impression on future low temperature research physicists for very many years.

Examining Kurt Mendelssohn's post-World War II work, it can be seen that he and his students continued to emphasize research on superconductors and liquid helium II.

A nice series of four papers on recent work in He II film flow, which also connected to earlier work in the Thirties, was published in the *Proceedings of the Physical Society* in 1950. Each paper had as authors Mendelssohn and a different student.

The first paper [2], written with John Daunt, describes work that had been done prior to the war but not yet published. These experiments involved more studies of the thermo-mechanical effect (Chap. 4) and film flow. The experiment is shown in Fig. 6.2. A vessel containing He II is inserted into another He II bath. The vessel contains a heater and is connected to the outer bath by a set of copper wires. The wires are present to increase the surface area over which the He II film may flow between the vessels. At the start, the helium levels in the vessel and the outer bath are the same (as discussed in Chap. 4). However, as heat is applied via the heater to the inner vessel, helium flows via a helium film from the outer bath to the inner vessel raising the helium level in the inner vessel. This is the thermo-mechanical effect.

What Mendelssohn's group observed was that while the change in level rose as more heat was added to the inner vessel, there was always a point at which the level

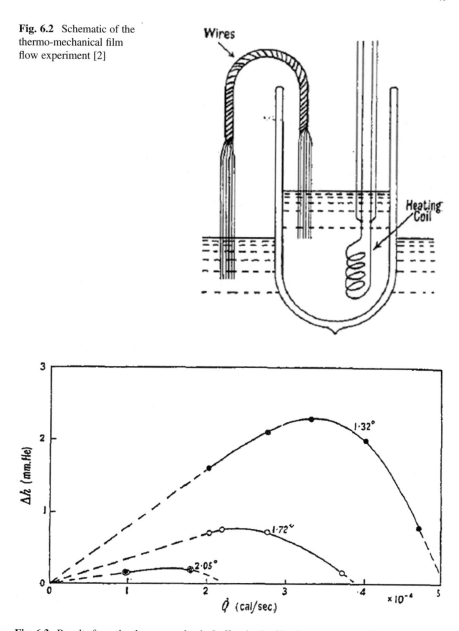

Fig. 6.2 Schematic of the thermo-mechanical film flow experiment [2]

Fig. 6.3 Results from the thermo-mechanical effect in the film flow experiment [2]

started to go back down with increasing heat (see Fig. 6.3). The interpretation of these results is that the velocity of the helium film flow reaches a critical velocity that it cannot exceed. The inner vessel level dropped due to increased evaporation of the liquid helium, resulting from the increased heat flow, which cannot be replaced due to the limited velocity of the film flow from the outer vessel. At this time, there was no good explanation for why such a critical velocity existed.

The second and third papers [3, 4] involved to a large extent experiments carried out in response to results from other research groups. K. R. Atkins from the University of Cambridge reported much higher rates of helium film flow than were seen by Kurt Mendelssohn. This seemed to be linked to the flow going through very small openings. Mendelssohn reproduced these experiments [3] in his laboratory but did not see these higher flows. At about the same time, W. J. de Haas and G. J. Van Den Berg from the University of Leiden also reported much higher film flow rates which they claimed were due to the blocking of thermal radiation from shining on the film. Dr. Mendelssohn and his team carefully designed an experiment [4] that compared the film flow rates in the presence and absence of thermal radiation. They did not see higher flows and observed no real difference between the cases in which thermal radiation was present or not.

However, the question remained, what explains the higher film flow rates observed in Cambridge and Leiden? Kurt Mendelssohn hit upon the idea of contamination. He proposed that the change in transfer rates might be due to gases such as components of air (oxygen, nitrogen, argon, neon etc.), water vapor or hydrogen freezing out on the surface of the vessels and affecting the film flow. Since literally everything but helium becomes a solid at liquid helium temperatures, freezing of contaminants is a common problem in cryogenic systems and great efforts are taken to avoid the presence of such contaminants.

Dr. Mendelssohn's team created an experiment in which they were able to introduce known amounts of gases, other than helium, into the system which froze out on the vessel walls. Now they observed the higher transfer rates seen at Cambridge and Leiden (see Fig. 6.4). The evidence seems clear, minute amounts

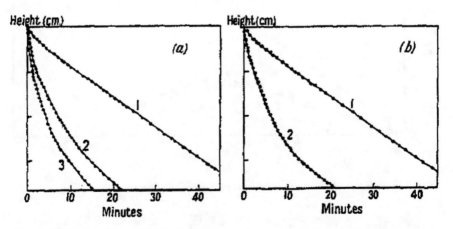

Fig. 6.4 (a) Rate of level height change in film flow as a function of surface contamination. Contamination with hydrogen: *1* Clean beaker, *2* First deposit, *3* Second and third (heavy) deposit [5]. (b) Rate of level height change in film flow as a function of surface contamination. Contamination with neon: *1* Clean beaker, *2* Neon deposit [5]

of impurities frozen out on the glass walls of the vessels caused the higher transfer rates. The physical mechanism for increase in transfer rate was unknown and Mendelssohn likely did not see this effect in his original work due to his typically careful experimental technique that avoided these impurities.

The careful study of film flow in liquid helium near absolute zero temperature may appear to only be of academic interest. However, the lower temperatures and unique heat transfer properties of He II have turned it into a vital coolant for many important large-scale scientific experiments. He II cools the superconducting magnets that make the Large Hadron Collider (LHC) at CERN possible. It has also become the standard coolant for superconducting radiofrequency accelerating cavities which are widely used in particle accelerators, such as those at the European Spallation Source, the European X-Ray Free Electron Laser, and the Linac Coherent Light Source II. The low temperatures of He II have made it important to infrared astronomy and it has been employed in a number of space-based experiments, including the Spitzer Space Telescope and the Herschel telescope. Since its discovery in the 1930s, He II has evolved from a laboratory curiosity to a well understood industrial coolant [6]. Mendelssohn's research on the fundamental behavior of He II, along with work at Cambridge, Moscow, Leiden and other institutions have helped make this possible.

Experiments in He II continued to be a topic of both papers and student dissertations throughout the rest of Mendelssohn's career.

Superconductivity work continued as well, although it is clear from the group's output that a larger emphasis was put on cryogenics over superconducting materials. Nevertheless, Mendelssohn and his team did stay current in the field, and work was carried out on the new Type II superconductors along with their related topic of flux pinning as these materials were developed in the 1960s.

Another area of significant research for Kurt Mendelssohn and his team was the properties of materials at liquid helium temperatures. This harks back to Mendelssohn's original thesis work in Berlin on the specific heat of solid hydrogen. The properties of thermal conductivity, electrical resistivity and specific heat were measured for a variety of materials including: cadmium, tantalum, germanium and niobium. Dr. Mendelssohn carried out some of this work with his student H. M. Rosenburg who later made his career in solid state physics.

Starting in 1957, a sequence of Mendelssohn's students investigated the properties of an unusual set of materials at cryogenic temperatures. These were the radioactive actinide elements, above all transuranic plutonium and neptunium. This work resulted from Kurt Mendelssohn's appointment as a part-time consultant at the U.K. Atomic Energy Research Establishment in Harwell which is not far from Oxford. He held this position along with his Oxford posts until his retirement.

The Harwell-related research came about this way. Firstly, in New Mexico at Los Alamos scientists had very recently taken their research on the highly radioactive alpha-plutonium metal down to the temperatures of liquid helium. At Harwell it was essential that their plutonium research team should do so too. To achieve this, Dr. Malcolm Waldron, head of the plutonium metallurgy department, and Dr. James A. Lee, head of the plutonium research team, approached Kurt

Fig. 6.5 Terence Meaden and his ³He cryostat at the Oxford Clarendon Laboratory 1962 (Courtesy of T. Meaden)

Mendelssohn to find a means by which such work could be launched at Harwell which till then had not cooled research specimens to temperatures lower than liquid nitrogen (77 K). A fitting application of Kurt Mendelssohn's long experience in working at liquid helium temperatures is what Harwell needed. He agreed. He would design a variation of a conventional cryostat that would be used at Harwell with an argon-filled glove-box suitable for studying highly perilous specimens. It would be built for Harwell in the Clarendon workshop, and a similar cryostat would be built that would remain in the Clarendon. A doctoral student would be found who would learn the cryogenic techniques at Oxford and transfer the knowledge of liquid helium cryogenics to the plutonium laboratory at Harwell.

In July 1957 Terence Meaden (Fig. 6.5), a finals year Oxford physics student who had just graduated, was selected to join the team. A major objective at Harwell would be to determine whether the artificially created actinide metals plutonium and neptunium (atomic numbers 94 and 93) were superconducting at low enough temperatures—and it was hoped that enough americium (atomic number 95) would become available too. There was small but cautious optimism for this because pure thorium (atomic number 90) becomes superconducting below 1.4 K and pure uranium (atomic number 92) at about 0.75 K. It was safe for these latter metals to be studied in Oxford. Electrical resistivity measurements would be done initially as well as thermo-electricity.

After studying the electrical resistivity of thorium and uranium from room temperature down to their superconducting levels, Meaden moved to Harwell to

Fig. 6.6 Cryostat
(projecting below at the left,
with dewar removed)
mounted in a glovebox at
AERE Harwell for
investigations of electrical
properties of actinide metals
(Courtesy T. Meaden)

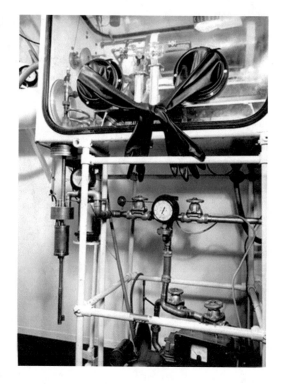

search for the onset of superconductivity in the intensely radioactive metals plutonium and neptunium but he found no evidence in either down to temperatures of
about 0.75 K [7]. Figure 6.6 shows Meaden's cryostat incorporated into a glovebox
at Harwell. At the time the hope continued that a sufficient quantity of the next
actinide metal in the series—the even rarer americium—would become available but
it never did. Fifteen years later J.L. Smith and R.G. Haire in the U.S. found
americium to be superconducting at and below 0.8 K. Another actinide metal,
protactinium (atomic number 91), is now known to be a superconductor at 1.4 K.

The change of electrical resistivity of plutonium with falling temperature is
remarkable for its extreme abnormality as can be judged by Fig. 6.7. That of
neptunium is unusual too (Fig. 6.8). Common among metals is a simple proportion-
ality of resistance with temperature.

This challenging work was published in the *Proceedings of the Royal Society* in
1963. Later, working at the Clarendon Laboratory with plutonium and neptunium
protected in sealed capsules, Dr. Meaden with Japanese colleague Dr. Shigi from
Osaka University reached 0.50 K for plutonium and 0.41 K for neptunium by means
of pressure reduction of liquid ^{3}He[1] (liquid helium-3) that Mendelssohn had sourced

---

[1]He II (see Chap. 4) is the second liquid phase of the most common isotope of helium known as ^{4}He
with the 4 representing the 2 protons and 2 neutrons in the helium nucleus. There is another, much
rarer, isotope ^{3}He with 2 protons and 1 neutron in its nucleus. ^{3}He becomes a superfluid when
cooled below 2.65 mK.

Fig. 6.7 The abnormal
temperature dependence of
the electrical resistance of
alpha-plutonium (at left) is
characterized by a negative
temperature coefficient
above 105 °K and a large
positive coefficient below
this temperature. Some
plutonium rich delta-Pu + Al
alloys showed similar
behaviour [8]

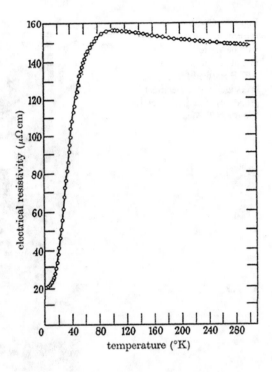

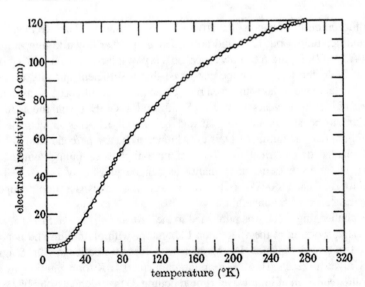

Fig. 6.8 The electrical resistivity of neptunium (at right) is unusual too [8]

in gaseous form. There was no superconductivity at these temperatures. The apparatus is described in an early issue of *Cryogenics* (1964) and an anecdotal summary on-line in a Newsletter of the Department of Physics, 2016.

Terence Meaden had become aware that when the same specimen of plutonium was cooled repeatedly from room temperature to helium-4 temperatures, its electrical resistivity at all temperatures had risen slightly, and that much of this was the result of spending several hours at liquid helium temperatures. Study of this phenomenon of self-irradiation was passed to the next student David Wigley (1962–1966) and several fine papers resulted [9]. Kurt Mendelssohn spread the news widely at conferences across the world. The explanation for this effect of rising resistivity is that the release of alpha particles due to the radioactive decay of plutonium causes radiation damage in the crystal structure of the metal. A similar effect was observed for neptunium. In due course, beginning in 1964 Charles Griffin took over research in this subject (until 1968) while Peter Sutcliffe (1964–1968) worked on the specific heat of plutonium. Related work on low temperature properties of the actinides continued with the fifth and sixth students Stephen Blow (1965–1968) and Michael Mortimer (1967–1972).

Making accurate measurements at very low temperatures with highly radioactive materials is just the sort of challenge that did appeal to Mendelssohn and the work with transuranic and actinide metals continued until the late 1960s resulting in six doctoral dissertations. All this later research on radioactive metals and alloys was done at A.E.R.E, Harwell, none at the Clarendon.

While it would be unfair to either generalize about Dr. Mendelssohn's students or to pick out specific students as typical, both these approaches need to be taken to understand his impact on his students and the impact of his students on science.

From 1934 to 1973 Dr. Mendelssohn supervised 50 doctoral students each of whom wrote a dissertation (see Appendix). All but two of these dissertations were on topics in cryogenics, liquid helium and superconductivity. The two exceptions: S.R. Reader and Henry M. Whyte, both of whom graduated in 1951, wrote dissertations on aspects of Medical Physics. These dissertations represent the completion of Dr. Mendelssohn's wartime activities. In addition to students, Mendelssohn hosted a number of visiting scholars who were also an integral part of his group.

There are a number of aspects known about Mendelssohn's research group. Given Kurt Mendelssohn's frequently demonstrated aversion to dealing with incompetent people, any student that graduated from his group was highly qualified in cryogenics. Further evidence of this is their publication record which is not only prolific but also frequently involved prestigious journals such as *Nature, Proceedings of the Royal Society* and *Physical Review*. Having come from a tradition in which researchers built their own equipment (even including liquefiers) Mendelssohn certainly taught practical aspects of cryogenics to his students. Such techniques would include construction of cryostats, leak detection, and wiring of instrumentation. Kurt Mendelssohn was very "hands on" in the laboratory, particularly in the early years at Oxford, working side by side with his students until the mid-1950s.

A frequently observed feature of the research by Mendelssohn's group is elegance of the experiments. A good example is the experiments developed to

Fig. 6.9 B. S. Chandrasekhar, K. Mendelssohn and D. Brewer (Courtesy of B. S. Chandrasekhar)

understand fundamental aspects of film flow in He II (see also Chap. 4). Kurt Mendelssohn was also known for taking care of his students. B. S. Chandrasekhar (Fig. 6.9), who graduated in 1952 remembers:

> I had an army surplus radio in my lab, and listened to music sometimes. Mendelssohn arrived sometime after midnight with sandwiches and a bottle of beer for me. We looked at the data and discussed what to do next. The music one time was Richard Wagner. He then told me of his visits as a student to Bayreuth to hear Wagner operas at the Festspielhaus. He took his own sandwiches, could not afford to buy them.
>
> Another of my memories of Mendelssohn speaks to the way he was with his students. I lived in my College, Queen's, on High Street. The main gate was locked from midnight to six in the morning. A late comer either climbed up a drainpipe on the High Street front and tiptoed through a don's room, or slept on a park bench somewhere. At the Clarendon Lab, I spent the day getting the experiment ready and producing about 200 mL of liquid helium, pumping it down below the lambda point leaving about 100 mL superfluid helium which evaporated at a rate of about 1 mL/h. This meant that I worked through the night and into the next, sometimes even longer. In between I took a short nap on a long table in the library a few doors away from my lab. Mendelssohn dropped in as usual late at night. If we decided that the experiment could be put on standby mode, he took me to his house on Iffley Road to sleep there the rest of the night. Breakfast with the family, and then back with him to the Clarendon and the experiment. Mendelssohn was partly continuing the tradition of some German professors in the twenties and thirties that put up their research students in their own homes.

Kurt Mendelssohn also assisted his students after they graduated by encouraging them to apply for positions, writing letters of recommendations and suggesting them for open job opportunities or as authors of books.

A somewhat more speculative observation on Dr. Mendelssohn's training of his students involves scientific communication. Given his emphasis on, and talent with, writing both for scientists and for the general public it is likely that he passed this interest on to his students. It is certainly clear that Kurt Mendelssohn encouraged his students to publish papers during their time in Oxford. It is also true that a number of Mendelssohn's students (see below) produced well-regarded books or founded scientific journals in their later careers.

A sense of the impact that Mendelssohn's group had on the field of cryogenics can be achieved by looking at the careers of four of his students and one of his visiting scholars.

John Daunt (1913–1987) was one of Dr. Mendelssohn's earliest students at Oxford. As described in Chap. 4, he started out working with Mendelssohn on properties of superconductors but then moved on to conduct pioneering studies of film flow in He II. Helium, superfluids, film flow and quantum mechanics would occupy most of Daunt's research career. After completing his D.Phil. in 1937, John Daunt stayed on at the Clarendon Lab as a research associate. In 1940, he became a Lecturer in Physics at Exeter College (Oxford University) and a Demonstrator in Physics at Oxford University. During the Second World War he worked on micro-wave RADAR and infrared physics; he recalls that one of his early papers with KM was on measuring the absorption of infrared light by superconductors.

After the war, John Daunt moved to Ohio State University becoming a full professor in 1950. In 1965, he went to the Stevens Institute of Technology where he founded the Center for Cryogenics. Prof Daunt ended his academic career at Queen's University in Canada. Over the years, a significant part of Daunt's research involved superfluid ^{3}He (including studies of ^{3}He films) and the properties of ^{3}He/^{4}He mixtures. He also continued his work in superconductivity. His research work was very productive and he won a number of awards and honors during his life.

John Daunt's academic life echoed Kurt Mendelssohn's in a number of ways. Like Mendelssohn, he travelled widely, with visiting professorships at Harvard, University of Amsterdam, Sao Paulo University and Columbia University among others. He is credited with helping to set up the Low Temperature Laboratory at Sao Paulo University.

John Daunt was the first editor of the *Journal of Low Temperature Physics* which started in 1969. Kurt Mendelssohn was on the first editorial advisory board of this journal. It is worth pointing out that two of the most important scientific journals in the field of cryogenics today were founded by Dr. Mendelssohn (*Cryogenics*—see Chap. 7) and by one of his students (*Journal of Low Temperature Physics*).

A later student, Harold Max Rosenberg (1922–1993), stayed at Oxford and did most of his professional work outside of cryogenics. Rosenberg graduated in 1953 and during his time in Mendelssohn's group studied the properties of materials at low temperatures. His thesis topic was "Energy Transport at Low Temperatures". He returned to Oxford University 6 years later as a University Demonstrator and later a

Reader. He was also a fellow at the University's Linacre College and St Catherine's College. Rosenberg's research career was in the area of solid-state physics including studies of metals, composites and amorphous materials along with the interaction of magnetism and quantized vibrations in solids [10]. He did pioneering work in applying fractal mathematics to the study of solid state physics.

While some of his research did involve low temperatures, much did not and Rosenberg is illustrative of a key feature of graduate education in science and engineering. Many, perhaps most, graduate students do not stay in the same subject area that they studied in graduate school. Funding opportunities, changing priorities in science, individual interest and other factors resulted in graduates working in other areas, sometimes even outside science and engineering. Thus, the impact of graduate student training in a group like Dr. Mendelssohn's will extend beyond cryogenics into other fields.

Rosenberg is also well known for writing two very good textbooks *Low Temperature Solid State Physics* and *The Solid State*. Just as in the case of Kurt Mendelssohn's books and perhaps learned from his association with him, Rosenberg's books are very clear and use graphical illustrations to explain complicated concepts.

Another example of Kurt Mendelssohn's students is Guy K. White (1925–2018) who completed his doctorate in 1950 with a thesis titled "Investigations on Liquid Helium". While at Oxford, he worked on He II film flow and the flow of He II through small capillaries. Originally Australian, Dr. White returned to Australia when he graduated, working at the Commonwealth Scientific and Industrial Research Organization's (CSIRO) National Standards Laboratory. He played an important role in developing the laboratory's cryogenic capabilities and conducted a wide range of experiments on the properties of materials at cryogenic temperatures. He remained at this laboratory during his career except for 4 years (1953–1954 and 1955–1958) at the National Research Council in Ottawa, Canada.

Just as Mendelssohn and Rosenberg did, Guy White wrote a book that has become a standard text in cryogenics: *Experimental Techniques in Low Temperature Physics* (1959). This book is now in its fourth edition.

Another student was G. Terence Meaden, the first of several who worked on the Harwell actinide projects. During his 6 years he spent some 3 years at A.E.R.E. Harwell and 3 years in the Clarendon. Upon leaving he pursued low temperature research at the Centre de Recherches sur les Très Basses Températures and the Department of Physics at the University of Grenoble in France, and then transferred to Dalhousie University in Halifax, Canada, as an associate professor of physics and was soon tenured. Low temperature research on metals and alloys, and the planning and building of cryogenic apparatus continued in the Mendelssohn style while employing a team of four doctoral students. Publications post-Oxford were numerous—54 on low temperature physics by 1975—in such journals as *Physical Review Letters (4)*, *Physical Review (1)*, *Cryogenics (15)*, *Journal of Low Temperature Physics (2)*, *Journal of the Less Common Metals (4)*, *Journal de Physique (2)*, *Metallurgical Reviews (1)*, *Journal of the Physical Society Japan (2)*, *Nuclear*

Metallurgy (1), *Physica Status Solidi* (2), *Physics Letters* (4), *Canadian Research and Development* (1), and *Journal of Thermal Analysis* (1).

Eventually, as expressed earlier for Harry Rosenberg, "changing priorities in science, individual interest and other factors" including health and family led Prof. Meaden to return to Europe to work in other realms, notably meteorological science (magazine editing, tornado climatology with dozens of papers, consultant to New Build Nuclear Power Stations) and archaeology which he had begun as a student at high school. This led to the archaeological solving by the scientific method of the meanings of dozens of prehistoric stone circles across Britain and Ireland. Above all, the core meaning of Stonehenge in the land of his Wiltshire ancestors was solved. This includes two papers in *The Journal of Lithic Studies* (Edinburgh University) and a summary in Wikipedia. It would have been in the nature and spirit of Kurt Mendelssohn as polymath to enjoy the turns and expansion of Meaden's scientific endeavours, having discussed at length with him the problems of the pyramids.

Professor Meaden also of course provided a preface to this book which gives a good sense of Mendelssohn as a person, detailing his dedication to both science and his students.

Kurt Mendelssohn also hosted a significant number of visiting researchers. These scientists, frequently from other countries, made contributions to the cryogenic program at Oxford and then moved to other institutions taking what they learned at Oxford with them. This is yet another means by which Mendelssohn's group influenced cryogenic research.

An example of such a visitor is Dr. Vinod Kumar Chopra (born 1942) Dr. Chopra graduated from Delhi University with a physics degree and then earned a PhD in Physics (specializing in physics at liquid helium temperatures) from the Indian Institute of Technology—Kanpur in 1970. Shortly after graduation, he was awarded the prestigious Nehru Memorial Trust Fellowship.[2] This fellowship enabled Dr. Chopra to study with Mendelssohn at Oxford and he worked there from 1970–1973. While there, Dr. Chopra was a Junior Research Fellow at Wolfson College, which by then was Mendelssohn's home college. During his time at Oxford, Chopra built a dilution refrigerator to reach milliKelvin temperatures and conducted research at both cryogenic temperatures and high magnetic fields.

Dr. Chopra has a memoir of his time in Oxford [11] describing both his research and his life in England. He describes that:

> Dr. Mendelssohn was known to be very helpful to the students, in particular the foreign visitors. He used to host large dinner parties at his house, for his students. The dinners at Dr. Mendelssohn's house were great fun. Dr. Mendelssohn and Mrs. Mendelssohn used to treat us as their own children & make us feel at home. Mrs. Mendelssohn would put a lot of extra effort to specially cook vegetables for the vegetarians among us.

---

[2]The importance of this Fellowship may be seen by the Chairs of the Fellowship Committee when it was awarded to Dr. Chopra. They were Lord Mountbatten of the UK and Prime Minister Indira Gandhi of India.

At the end of his time in Oxford, Dr. Chopra started to apply for positions at Indian research institutes. When Mendelssohn heard about this he exclaimed "Did not you know that my students do not apply; they get invited for the jobs!" [11]. Kurt Mendelssohn also pointed out that given his experience in England, Chopra would likely be much happier in the more Westernized city of Bombay (Mumbai) than say in Delhi. Mendelssohn then wrote to the Director of the Bhabha Atomic Research Center (BARC) in Bombay suggesting Dr. Chopra for a position. This institute was part of the complex of institutes that had impressed Dr. Mendelssohn during his India trip at the end of 1969 (Chap. 8). BARC hired Dr. Chopra, who established a new cryogenics laboratory and liquid helium facility. Mendelssohn also saw to it that Dr. Chopra was able to take back to BARC the dilution refrigerator that he had built in Oxford. As Dr. Mendelssohn was retiring at about the same time Dr. Chopra returned to India, he also arranged for Chopra to take back with him some other surplus equipment from Mendelssohn's lab. This equipment helped greatly in the setup of the new cryogenics laboratory at BARC.

Dr. Chopra stayed at BARC until 1982, after which he joined Goodwill Cryogenics which was an early supplier of cryogenic components to India.

In summary, Dr. Mendelssohn's research group was strong and made important fundamental discoveries in helium II, superconductivity and properties of materials at cryogenic temperatures. His students, upon graduation, continued to conduct important research in cryogenics and other fields. They also helped establish new centers for cryogenics in locations as diverse at Sao Paulo, Bombay and Sydney. Tellingly, his students made significant contributions to the literature of cryogenics. He assisted his former students and visitors with letters of recommendation, advice and in some cases equipment. He continued to collaborate with many of them professionally after they left Oxford.

Kurt Mendelssohn's research was recognized by a number of awards (see Epilogue). Dr. Mendelssohn was elected a Fellow of the Royal Society in 1951. His research and teaching works were acknowledged by Oxford University when it gave him permanent positions as a University Demonstrator in 1947 and as a Reader[3] in the Physics Department in 1955.

After his election to the Royal Society Kurt and Jutta Mendelssohn attended the annual Royal Society dinner in the winter and Conversazione in the summer. The dress code was formal white tie (Fig. 6.10) and suitably attired they drove off every year to the Dorchester Hotel in London. Unfortunately, on one occasion the car broke down and having left it with a garage KM decided that the only possible way of getting to the Dorchester in time was to hitch hike. They had not been standing by the roadside for long when a Rolls Royce drew up and the chauffeur asked them their destination and having received the information the chauffeur indicated that they

---

[3] At this time in Oxford, there were very few people with the title of Professor, mainly heads of Institutes and holders of named chairs. A Reader, being a permanent, full time staff member engaged in research and teaching would be the equivalent of a full Professor at any other institution.

Fig. 6.10 Kurt and Jutta
Mendelssohn at the Royal
Society (Courtesy
M. Mendelssohn)

should get into the car and off they smoothly drove to arrive at the Dorchester Hotel in time.

An example of the importance of Mendelssohn's group in cryogenics can be seen in an interview that G. K. White gave later in his life in which he described why he went to Oxford: "Oxford at that time was the key low temperature laboratory. Cambridge had a good one but it was much smaller... But Oxford was the key place. Partly because they got three guys I admired very much who all left Berlin" [12].[4]

References

1. F.E. Simon et al. *Low Temperature Physics: Four Lectures*, (Pergamon Press, London 1952)
2. J. G. Daunt and K. Mendelssohn, "Film Transfer in He II: I—The Thermo—Mechanical Effect", Proc. Phys. Soc. A 63 1305 (1950).
3. J. B. Brown and K. Mendelssohn, "Film Transfer in He II: II—Influence of Geometrical Form and Temperature Gradient", Proc. Phys. Soc. A 63 1312 (1950).
4. R. Bowers and K. Mendelssohn, "Film Transfer in He II: III—Influence of Radiation and Impurities", Proc. Phys. Soc. A 63 1318 (1950).
5. G.K. White and K. Mendelssohn, "Film Transfer in He II: IV—The Transfer Rates on Glass and Metals", Proc. Phys. Soc. A 63 1305 (1950).
6. J. G. Weisend II, "He II: From the Lab to an Engineering Fluid", Cold Facts, April (2014).
7. J.A. Lee, G.T. Meaden and K. Mendelssohn. "Low Temperature Resistivity of Plutonium and Neptunium". Proc. Phys. Soc. LXXIV, 671 (1959).

---

[4]Here White is referring to Francis Simon, Kurt Mendelssohn and Nicholas Kurti.

8. G.T. Meaden. "Electronic Properties of Actinide Metals at Low Temperature", Proc. Roy. Soc., 276:1367 (1963).
9. C. S. Griffin, K Mendelssohn, M. J. Mortimer, "Self-Irradiation Damage in the Actinide Metals", Cryogenics, April (1968).
10. "Obituary—H. M. Rosenberg", F. N.H. Robinson, *Cryogenics* Vol 34 (1994).
11. V. Chopra, Oxford Days, (in preparation)
12. "Dr. Guy White (1925–2018) Physicist" Interviewed by Prof N. Fletcher, Australian Academy of Sciences (2010)

Chapter 7
Connections: 1945–1980

"The increasing importance of this field, particularly in view of the missile development and low temperature electronics of all kinds as well as the progress in the main field of liquefaction and separation seem to justify fully such a journal"
K. Mendelssohn to J. B. Gardner in 1959 *regarding the founding of* Cryogenics

One result of World War II was the movement of physics from small laboratories to large industrial settings. The effort to build the atomic bomb (in which Francis Simon and Nicholas Kurti were involved) resulted in a complex of industrial sites in the USA and Britain employing thousands of people in the production of nuclear material and the design, production and testing of the bombs. Similar large-scale industrialization of physics research occurred in the fields of RADAR, communications and aviation. In addition to physics being applied to large scale industrial and defense applications, physics research itself also became large scale. Physicists using project management and other industrial skills learnt during the war effort began to develop large particle accelerators, telescopes and other facilities to conduct basic research. The view that science had helped win the war meant that there was a lot of funding available for physical science and engineering research and development. The era of "Big Science" had begun.

Kurt Mendelssohn recognized the new trend and was very interested in making the connection between cryogenics in the laboratory to cryogenics in service to industry, defense and big science. He did this by helping to found both the journal *Cryogenics* and the International Cryogenics Engineering Conference, co-editing the International Cryogenics Monograph book series, through his writings (Chap. 9) and by training many students who went to work in the cryogenics industry (Chap. 6).

He was also very interested connecting his laboratory at Oxford with universities in developing countries, to help them train students and develop research programs in cryogenics and superconductivity. These efforts also had the advantage of helping to satisfy his love of travel.

© The Author(s), under exclusive license to Springer Nature Switzerland AG 2021
J. G. Weisend II, G. T. Meaden, *Going for Cold*, Springer Biographies,
https://doi.org/10.1007/978-3-030-61199-6_7

There was another type of connection that Kurt Mendelssohn championed. This was the popularization of science. He did this through books, lectures and his work with the media.

The birth of the journal *Cryogenics* began in 1959. Dr. Mendelssohn was the editor of *Progress in Cryogenics,* a volume of scientific papers on topics in both cryogenic research and engineering. This would eventually become a four-volume set published between 1959 and 1964. The papers in these volumes were all designed to be summary papers of the current state of the art of various projects using cryogenics. These volumes illustrate two consistent features of Kurt Mendelssohn's efforts in scientific communication. First, the volumes tried to gather in one place overviews of topics in the increasingly disparate cryogenics community. Second, the papers were international in scope. Summation and international communication were hallmarks of all of his efforts in scientific organization and communication.

In late 1959, the publisher of *Progress in Cryogenics* (Heywood and Company of the United Kingdom) approached Kurt Mendelssohn with the idea of starting an international journal that would bring together papers on all topics in cryogenics in one place. Mendelssohn agreed that papers in cryogenics were then spread among many different journals and thought that such a publication was particularly timely due to the growth of the field.

The original plan was to have an Editor in Chief (Kurt Mendelssohn) assisted by three regional editors (one each from the USA, Continental Europe and the USSR) and to publish at first a quarterly journal perhaps moving to a bimonthly schedule depending on interest.

From the very beginning, the emphasis was seen to be on cryogenic engineering with papers on pure academic research taking a lesser role. Mendelssohn worked both to select regional editors who had a strong interest in cryogenic engineering as well as to solicit papers on engineering topics for the first issue.

The first issue of *Cryogenics* was published in September 1960 with Russell Scott of the Cryogenic Engineering Laboratory in Boulder, Colorado, USA and Louis Weil of the Low Temperature Laboratory at Grenoble University, France, assisting Mendelssohn as regional editors. The Editor-in-Chief idea seems to have been dropped and an editor from the USSR was not found. Both Scott and Weil's work was centered on cryogenic engineering. The first issue contained ten papers including ones on a Joule–Thompson helium liquefier, improvements in cryogenic insulation, and metallography at liquid helium temperatures. Three of the papers covered topics associated with ionizing radiation and cryogenics neatly illustrating the era's fascination with the uses of atomic energy. The issue also contained several shorter Letters to the Editors on technical subjects as well as a bibliography of other recently published papers on cryogenic topics in other journals. The goal was to try to bring together as much information as possible on cryogenics in one place. As part of the international character of the journal, *Cryogenics* also contained abstracts of the published articles in French, Russian and German.

Cryogenics became a bimonthly publication in 1964 and a monthly publication in 1973. The abstracts in languages other than English were eventually dropped for space reasons and the bibliography was dropped once other better lists of recent publications became available.

Today, *Cryogenics* is published monthly by Elsevier and is acknowledged as the principal international journal of cryogenic engineering. Publications such as *The Journal of Low Temperature Physics*[1] tend to publish papers on "pure" cryogenic research. In addition to its stress on cryogenic engineering, other aspects of the original plan for the journal remain. There are still three regional editors: one from Asia, one from Europe and one from North America.

Another significant method of communication is the scientific conference. Here, scientists would meet and present the results of their research. The value of such conferences goes well beyond the presentation of results. In fact, the greater value comes from the personal connections that are made between attendees. Such connections may result in new research collaborations, visiting professorships, job placement for graduating students or inspire new research directions. An example of such of conference is shown in Fig. 7.1. This photograph shows the attendees of the Lorentz-Kamerlingh Onnes Centenary Conference on Electron Physics held June 22–27, 1953 in Leiden, The Netherlands. Kurt Mendelssohn attended and presented a survey paper on "Thermal Conductivity of Superconductors" [1]. Note that in addition to Kurt Mendelssohn there are many other famous physicists: Niels Bohr, Paul A. M. Dirac, Werner Heisenberg, Brian Pippard, Rudolf E. Peierls, Fritz London, and Wolfgang Pauli.

The International Cryogenic Engineering Conference (ICEC) was founded as a response to the Cryogenic Engineering Conference (CEC) that had begun in the USA. The origins of the CEC can, in turn, be directly linked to the development of the hydrogen bomb.

The early hydrogen bombs developed after World War II used liquid hydrogen. Hydrogen gas becomes a liquid at 20 K at atmospheric pressure, and the liquefaction, storage and transport of liquid hydrogen requires cryogenic engineering. In order to meet the need for liquid hydrogen and the associated expertise to handle it, the US Atomic Energy Commission (AEC), responsible for the hydrogen bomb program, gave the US National Bureau of Standards funding to set up a cryogenic research and engineering laboratory in Boulder, Colorado. The efforts of this laboratory contributed directly to the first successful hydrogen bomb test in 1951.

By 1954, weapons designers had found a way to store the hydrogen in the form of a room-temperature compound of solid deuterium[2] and lithium. This removed the need for cryogenics and the AEC greatly reduced funding for the Boulder laboratory. Desperate for funds and now able to conduct unclassified research, the staff at the Cryogenic Engineering Lab (CEL) at Boulder decided to hold an open conference to showcase their capabilities. This first Cryogenic Engineering Conference was held in

---

[1]This journal was started in 1969 by Plenum Publishing with John Daunt, a former student of Mendelssohn's (see Chap. 6), as its editor. Kurt Mendelssohn was on the original editorial board.

[2]Deuterium is an isotope of hydrogen containing one proton and one neutron rather than the single proton found in "regular" hydrogen.

Fig. 7.1 Attendees of Lorentz-Kamerlingh Onnes Centenary Conference on Electron Physics held June 22–27, 1953 in Leiden (Courtesy of Kamerlingh Onnes laboratory)

Participants in the Lorentz Kamerlingh Onnes Memorial Conference in Leiden, 1953. Standing, left to right: A.N. Gerritsen, A.F. van Itterbeek, H. Wergeland, J. Vlieger, K.W. Taconis, G. Borelius, A. Pais, M.J. Druyvesteyn, P.A.M. Dirac, R. de Laer Kronig, K. Mendelssohn, C.F. Squire, W.J. de Haas, L.J.F. Broer, J. de Boer, A.D. Fokker, W. Heisenberg, A.B. Pippard, N. Bohr, W.E. Lamb, D. Shoenberg, H.B.G. Casimir, J. Korringa, J. Dingle, R.E. Peierls, S.A. Wouthuysen, F.J. Belinfante, G. Källén, J. Smit, H. Fröhlich, J. de Nobel, F. Bloch, F. London, H.J. Groenewold, G.W. Rathenau, P.H.E. Meyer, J.G. Daunt, D.K.C. MacDonald, S.R. de Groot, R.M.F. Houtappel, B.R.A. Nijboer, W. Pauli, L. Rosenfeld. Sitting from left: J. Clay, M.A. Proca, S. Tomonaga, H.M. Gijsman, C.J. Gorter, B. Ferretti, J. van den Handel, G.J. van den Berg, M. Fierz.

Boulder in September 1954 with 400 participants. The CEC continues to be held biannually in North America and is one of the two major conferences in cryogenic engineering.

The other major cryogenic engineering conference is the International Cryogenic Engineering Conference (ICEC) founded in part by Dr. Mendelssohn. The continuing success of the Cryogenic Engineering Conference along with the growth of the US cryogenics industry (funded to a large extent by the space program) worried Mendelssohn. He was concerned that scientists and engineers outside of the USA would miss out on the development of this technology and that as a result the cryogenics industry in Europe would lag behind. At about the same time, the Cryogenic Association of Japan became interested in holding an international cryogenic conference in Japan. An ad hoc international committee was formed to organize the conference with Dr. Mendelssohn as the chair and Prof. K. Oshima of the University of Tokyo as its secretary. Other committee members included Russell Scott from CEL, Boulder and Louis Weil (Dr. Mendelssohn's co-editors from *Cryogenics*) and I. P. Vischnev from the USSR. ICEC 1 was held in 1967 in Tokyo and Kyoto. There were 274 attendees for this first conference, mostly from Japan but with sizable contingents from the USA and Europe. Kurt Mendelssohn gave a talk on "Cryogenics: The Present and the Future" In this lecture, he touched on themes that he would consistently return to during his career: "Cryogenic engineering, as it faces us today, requires many and diverse skills and disciplines. So far, they often still stand apart but they will have to cooperate and enter upon each other's field if they wish to get results … It must be our aim to further this cooperation by holding meetings, such as the present one, at which people from all these different fields will meet to learn and speak the common new language of cryogenic engineering" [1].

The success of this meeting led to the holding of ICEC 2 in Brighton, UK, in 1968. It was during this meeting, that it was decided to form a permanent International Committee to oversee the organization of a series of biannual International Cryogenic Engineering Conferences. During these discussions, Dr. Mendelssohn made it clear that he did not want the ICEC conferences to "shut out any cooperation with their American colleagues". He was also very interested in involving researchers from the USSR. A caretaker committee with Kurt Mendelssohn as the chair was established and the next conference was planned for West Berlin in 1970. During his opening address at this conference he reiterated the committee's desire not to exclude American participation (there were Americans present both on the International committee and in attendance at the conference) and to bring together workers in different disciplines interested in cryogenics.

The International Cryogenics Engineering Conference continues to this day. It still looks very much like the original vision of its founders. The conference is held biannually with the location generally alternating between Europe and Asia. In order not to conflict with the Cryogenic Engineering Conferences in North America, the ICEC is traditionally held in even years and the CEC is traditionally held in odd years. There is still US participation in the ICEC conferences as well as significant non-US participation in the CEC conferences.

Table 7.1 Mendelssohn Award Winners

Year of Award	Name	Country
1986	K. Oshima	Japan
1988	P. Roubeau	France
1990	J. Gardner	UK
1992	P. Mason and D. Petrac	USA
1994	H. Desportes	France
1996	O. Lounasmaa	Finland
1998	V. Arp and R.D. McCarty	USA
2000	C. S. Hong	China
2002	G. Claudet	France
2004	I. and G. Klipping	Germany
2006	P. Komarek	Germany
2008	H. Quack	Germany
2010	S. W. Van Sciver	USA
2012	Sir Martin Wood	UK
2014	R. Scurlock	UK
2016	Ph. Lebrun	Switzerland
2018	A.T.A.M. de Waele	The Netherlands
2020	G. Gistau Baguer	France

Dr. Mendelssohn participated in and championed the ICEC for the rest of his career. In 1986, the International Cryogenic Engineering Conference established a lifetime achievement award and named it in his honor. The Mendelssohn Award recognizes people who make:

- "New and promising solutions to difficult problems
- Promotion and encouragement of work in new fields of low temperature applications, for stimulating the cryogenic community's interest in these fields and for helping to establish them
- Long-standing contributions to cryogenics" [2]

Table 7.1 lists the Mendelssohn Award winners to date.

The exchange of information among workers in cryogenics was also the motivation behind the International Cryogenics Monograph book series. This series resulted from discussions between Prof. K. Timmerhaus of the University of Colorado and K. Mendelssohn. Both believed that there was lack of in-depth books by experts on topics in cryogenics. Such books would complement the shorter, possibly more temporary papers, published in *Cryogenics* or in the proceedings of the CEC and ICEC. The books could also serve as text books for cryogenic engineering courses. Prof. Timmerhaus as the editor of *Advances in Cryogenic Engineering* (the Proceedings of the Cryogenic Engineering Conference) had contacts at Plenum Press who agreed to publish the series.

Prof. Timmerhaus and Dr. Mendelssohn served as co-editors for this series. From the beginning, the series was expected to involve authors from around the world.

Timmerhaus sought and worked with authors in the United States, while Mendelssohn did the same for authors in Europe. Editorial decisions were made jointly by Mendelssohn and Timmerhaus. Finding suitable, qualified authors for such a book series is always a challenge and both Mendelssohn's and Timmerhaus' contacts within the cryogenics community were invaluable in this effort.

A total of 13 books were published in this series before Kurt Mendelssohn's death.

The International Cryogenic Monograph Series continues today. It is now published by Springer and edited by Sangkwon Jeong (KAIST, South Korea) and J. G. Weisend II (European Spallation Source, Sweden). The series remains a source of high-quality technical books on cryogenics. Table 7.2 shows all the books in the series to date.

The explanation of science to the general public was a favorite topic of Kurt Mendelssohn's (Fig. 7.2). All his books, even those on technical subjects, were aimed at the general reader. He talked to the press and was interviewed a number of times on BBC radio concerning cryogenics, his travels, science in developing countries and his books. These broadcasts, which were often reprinted in the BBC publication *The Listener*, would many times result in letters being sent to him. He would take the time to read and answer these letters, frequently suggesting other sources of information to the writer. Even when the letter betrayed some ignorance of the topic, Mendelssohn would respond politely and correct the writer. Not every letter was answered. He kept a "nuttier" file of really outrageous letters (ones claiming alien spacecraft as an explanation for physical phenomena, for example). These tended not to be answered.

Dr. Mendelssohn also wrote book reviews and letters to the editor for various publications. He facilitated the filming of scientific demonstrations in his laboratory at Oxford for a television program.

During his career at Oxford, in addition to his university lectures, he taught a number of lectures in both low temperature and general physics for the Extra-Mural Studies Program (essentially a night school for adult students at Oxford). He also gave lectures to student groups such as one in 1976 to the Chemical and Physical Society of the University College London on his activities in Berlin in the twenties.

There are probably several reasons for Kurt Mendelssohn's interest in popularizing science. When it comes to his books, it may partly have been a marketing move; the broader the audience, the more sales. It may hark back to his childhood in Germany where the attending of public lectures on science and technology was, and still is, seen as a viable entertainment option. It may also have to do with his occasional lament about bright people going into business rather than science and his desire to change this situation. However, the real reason may be the simplest. Kurt Mendelssohn enjoyed figuring things out and once having done so enjoyed explaining them to others.

Table 7.2 Books in The International Cryogenic Monograph Series

Author/Editor(s)	Title	Year
G. Gistau Baguer	Cryogenic Helium Refrigeration for Middle and Large Power	2020
Atrey, Milind (Ed.)	Cryocoolers: Theory and Applications	2020
Scurlock, Ralph & Bostock, Thom	Low-Loss Storage and Handling of Cryogenic Liquids—The Application of Cryogenic Fluid Dynamics, Second Edition	2019
Weisend, John; Peterson, Thomas	Cryogenic Safety: A Guide for the Laboratory & Industry	2019
Jacobsen, Richard T., Penoncello, Steven G., Lemmon, Eric W., Leachman, J, Blackham, T.	Thermodynamic Properties of Cryogenic Fluids—2nd Edition	2017
Weisend, John (Ed.)	Cryostat Design: Case Studies, Principles and Engineering	2016
Ventura, Guglielmo, Perfetti, Mauro	Thermal Properties of Solids at Room and Cryogenic Temperatures	2014
Maytal, Ben-Zion, Pfotenhauer, John M.	Miniature Joule-Thomson Cryocooling: Principles and Practice	2013
Pavese, Franco, Molinar Min Beciet, Gianfranco	Modern Gas-Based Temperature and Pressure Measurements	2013
Van Sciver, Steven W.	Helium Cryogenics	2012
Venkatarathnam, Gadhiraju	Cryogenic Mixed Refrigerant Processes	2008
Timmerhaus, Klaus D.; Reed, Richard P. (Eds.)	Cryogenic Engineering: Fifty Years of Progress	2007
Barron, T.H.K., White, G.K.	Heat Capacity and Thermal Expansion at Low Temperatures	1999
Ackermann, Robert A.	Cryogenic Regenerative Heat Exchangers	1997
Jacobsen, Richard T., Penoncello, Steven G., Lemmon, Eric W.	Thermodynamic Properties of Cryogenic Fluids	1997
Edeskuty, Frederick J., Stewart, Walter F.	Safety in the Handling of Cryogenic Fluids	1996
Hartwig, Gunther	Polymer Properties at Room and Cryogenic Temperatures	1995
Timmerhaus, Klaus D., Flynn, Thomas M.	Cryogenic Process Engineering	1989
Collings, E.W.	Applied Superconductivity, Metallurgy, and Physics of Titanium Alloys: Volume 2: Applications	1986
Collings, E.W.	Applied Superconductivity, Metallurgy, and Physics of Titanium Alloys: Fundamentals Alloy Superconductors: Their Metallurgical, Physical, and Magnetic-Mixed-State Properties	1986
Walker, Graham	Cryocoolers: Part 1: Fundamentals	1983
Frost, Walter	Heat Transfer at Low Temperatures	1975
Savitskii, E M, Baron, V V, Eftimov, Yu V Bychkova, M I and Myzenkova, L F	Superconducting Materials	1973
Hurd, C M	The Hall Effect in Metals and Alloys	1972
Bailey, C A	Advanced Cryogenics	1971

(continued)

Table 7.2 (continued)

Author/Editor(s)	Title	Year
Wigley, David A	Mechanical Properties of Materials at Low Temperatures	1971
Croft, A J	Cryogenic Laboratory Equipment	1970
Smith, A U	Current Trends in Cryobiology	1970
Keller, W E	Helium-3 and Helium-4	1969
Parkinson, D H and Mulhall, B E	The Generation of High Magnetic Fields	1967
Zabetakis, M G	Safety with Cryogenic Fluids	1967
Gopal, E S R	Specific Heats and Low Temperatures	1966
Meaden, G Terence	Electrical Resistance of Metals	1965
Goldsmid, H J	Thermoelectric Refrigeration	1964

Fig. 7.2 In an updated photo Kurt Mendelssohn presents a public lecture and demonstration on science (Courtesy J. Mendelssohn)

References

1. K. Mendelssohn, "Cryogenics, The Present and the Future", Proceedings first International Cryogenic Engineering Conference (1967).
2. http://icec.web.cern.ch/content/mendelssohn-award Accessed July 2020.

Chapter 8
Travel: 1945–1980

"It is not inconceivable that historians of the future may consider the emergence of China as a major industrial power to be the most important development of the second half of our century"
K. Mendelssohn, Science *Sept. 22, 1961*

Kurt Mendelssohn loved to travel. It was an activity that brought together a number of passions: his desire to learn about different cultures and history, his enjoyment at making personal contacts with other scientists and his interest in connecting the research program at Oxford with the wider world. The list of major trips that he took includes places such as India, China, Japan, Ghana, Pakistan, Portugal, the USSR, the USA and Egypt. Added to this would be scores of shorter trips for scientific conferences, meetings and laboratory visits. Dr. Mendelssohn showed a particular interest in visiting and learning the culture and the state of scientific research in developing nations.

Two aspects of his travels stand out. First, his trips were very efficient. He generally developed, and seemed to enjoy, complicated routings that would allow him to visit as many places as possible on a single journey. For example, in the spring of 1960, he made a 2-month trip to Japan, but on the way stopped off in India and Thailand and on the way back visited Hong Kong, Beijing (his first time in China) and Moscow. These tours were efficient in another sense. He used them to accomplish a number of parallel goals. On a single excursion, he might make contacts with foreign scientists and universities, develop plans for collaborative research, collect material that would later become a book, article or broadcast, recruit papers or assistance for *Cryogenics* or the International Cryogenic Engineering Conference (ICEC), and indulge his hobbies of photography and the collection of Chinese ceramics. True to form, he did a lot of preplanning: contacting people who knew about the place he was visiting, arranging for letters of introduction and setting up appointments before he left.

The second aspect of Mendelssohn's travels that stands out is his ability to find funding for them. He was adept at finding scientific societies or government agencies, both in the UK and abroad, to pay for his travel. When necessary, he would

© The Author(s), under exclusive license to Springer Nature Switzerland AG 2021
J. G. Weisend II, G. T. Meaden, *Going for Cold*, Springer Biographies,
https://doi.org/10.1007/978-3-030-61199-6_8

combine funding sources. This skill allowed him to take longer and more frequent trips. Some of the organizations that funded his travels over the years included The Royal Society, The Japanese Society for the Promotion of Science and The Soviet Academy of Sciences. His wife, Jutta, frequently accompanied him on these travels.

Given the extent of Kurt Mendelssohn's travels, it is best to examine in detail three of his journeys; his trips to Russia, India, and China.

In July 1957, he went to Moscow for 10 days to attend a low temperature conference along with five other British physicists. These were the only foreign attendees at the conference and they had been invited (though not funded) by the Soviet Academy of Sciences. While this was a relatively short visit, it marked his first visit to the USSR and there were several themes developed during this visit that are also seen on his later travels. Most notably, Kurt Mendelssohn wrote a short essay on his impressions from the trip that was published in the Bulletin of the Institute of Physics ("Moscow Journey" October 1957). Rather than being a dry recitation of the conference or the papers presented, this was really a popular travelogue that foreshadows his later writings on India, and particularly China. The essay contains a number of colorful descriptions: Prague airport is described as "that curious half way house between two worlds"; Moscow as the "most fascinating boom town of our time" and the pavilions in the Soviet agricultural and industrial exhibition as "unique architectural horrors" [1]. His definite opinions and his ability to use interesting language is a large part of what made his writings accessible and interesting to the general public.

The conference was hosted by the Institute for Physical Problems led by Prof. P. Kapitza (Figs. 8.1 and 8.2). It will be recalled that it was Kapitza's laboratory at Cambridge University that Mendelssohn beat in the effort to be the first to liquefy helium in the UK. He was the only member of the British delegation that had known Kapitza in England and was quite pleased to meet up with him again.

One topic that Dr. Mendelssohn comments on here and returns to frequently in his writings is the treatment of scientists. He was quite impressed with the resources available to scientists in the USSR. "Scientific ingenuity is evidently deemed the most precious commodity and is treated as such by both paying for it in high salaries and by providing good facilities for scientific work" [1]. Over his career, he consistently argued that science and scientists need the same level of status, resources and compensation as afforded to people in business or politics. One also gets the sense that he never thought that he, despite all his accomplishments—nor anyone else in leading-edge scientific research—was ever suitably compensated by any British Government.

During the Sixties, Kurt Mendelssohn made a number of trips to India and China and worked as a visiting professor in Ghana. India and China in particular fascinated him. It was not just that these were exotic lands with ancient civilizations, though Mendelssohn certainly was interested in their history and culture. Rather, India and China were just beginning to develop modern scientific laboratories and industries. Both countries had also only recently thrown off Western dominance. How these countries created scientific infrastructures and the role of scientists in the development of the countries were subjects that greatly interested him and ones on which he

Fig. 8.1 Kurt Mendelssohn (standing) in Russia. Lev Landau is seated at Kurt Mendelssohn's right. Piortr Kapitsa is seated to the right of Laudau (Courtesy of J. Mendelssohn)

had definite opinions. The majority of his writing and speaking about his travels in India and China dealt with these subjects.

Tellingly, Dr. Mendelssohn did not look down on scientists in developing countries nor did he automatically insist that they simply copy the West. Instead, he goes out of his way in his writings to compliment the creativity and intelligence of many of the scientists he met and frequently suggested that they need to adapt to local conditions to make progress. He certainly complained about bureaucracy, scientists that he felt were simply lazy or incompetent, and politicians who did not value science. However, in general, he gives developing countries such as China and India the benefit of the doubt. His first visit to India came in 1963 when he spent 5 weeks touring various research institutes and universities as well as the standard tourist sites. As he had just visited China a few months earlier, he was able to compare the state of scientific development of both countries. This comparison was even more relevant due to the recent Sino–India War in which China defeated Indian forces, pushed into Indian Territory and then retreated, leaving India in a state of shock.

Kurt Mendelssohn described his visit in a talk on BBC radio (later reprinted in *The Listener*) titled "Science in India". He explained that due to the recent war almost all the Indian universities and research centers had undergone a budget cut of twenty per cent and the result was a very disheartened research community. "Far

Fig. 8.2 Kurt Mendelssohn
in conversation with
Russian theoretical physicist
Lev Landau (Courtesy of
J. Mendelssohn)

from playing their part in helping their country, Indian scientists had been made to feel that in the national emergency money spent on science was ill spent and that they were not a help but a hindrance" [2]. Having experienced World War II where science, particularly physics had played such a large role in the victory, Mendelssohn of course thought the Indian government priorities were exactly wrong. He attributed this error to the fact that Indian politicians had developed their views on science before World War II when the British also did not value science as a necessity. He also said part of the problem was that the Indian Universities did not themselves value science, putting a higher value on law and the liberal arts. The result, according to Dr. Mendelssohn, was university science departments that were lethargic, underfunded and poorly staffed.

He said that ironically, since China has only recently started to adopt modern science, it learned from the examples of the Soviet Union and the West and gave science its proper due. According to Dr. Mendelssohn, "Mr. Mao Tse-tung had left me in no doubt about what the Chinese think. He emphasized China's urgent need for the expansion of science and technology" [2].

Kurt Mendelssohn places the blame for problems with Indian science on policies and politicians not on any innate problems with Indian culture. As an example, regarding the possibility of religion hindering science he says "I know of too many excellent Indian scientists who are also devout Brahmins to take this argument seriously" [2]. He felt that India possessed many highly skilled scientists and had a large reservoir of intellectual talent.

When it came to industrial development, Dr. Mendelssohn also thought that the Indian policy of basically importing complete factories and running them was flawed. He felt it would be better for these plants to be developed in India using Indian resources to develop its own national technical capabilities. He observed "It is not the know-how of science that Indian professors need as much as the know-how of talking to their government" [2].

At the end of his trip, he had a meeting with Prime Minister Nehru of India and described his ideas on Indian science and technology. He was then invited to give two talks to Members of the Indian Parliament and Civil Service contrasting science in China and India.

During his first trip to India, he did find several research institutes that he felt were doing high quality work. The one that impressed him most was the Tata Institute of Fundamental Research in Bombay (now Mumbai). It is thus not surprising that when he returned to India in 1969–1970 for a 4-month stay as a visiting professor, he went to the Tata Institute.

This was funded by the Royal Society, and on his return he produced a written report on his trip. Jutta accompanied her husband on this visit and he spent his time lecturing and carrying out research as well as making extensive tours of Indian Universities and research institutes.

Dr. Mendelssohn gave a series of general and then more detailed lectures on cryogenics while at Tata. Also, at the end of his stay, he gave well attended formal lectures on science in China and his observations on pyramids in Egypt and Mexico. This last topic shows that Mendelssohn was thinking about issues that he would later develop in his book *The Riddle of the Pyramids* (1974). His research work at Tata centered on experimental measurements of transport properties of liquid helium and were designed to complement the work he was currently doing in Oxford. He was quite pleased with the conditions, both professional and social, at the Tata Institute. He felt well taken care of, was impressed by the staff and had everything he needed to carry out his work. This last point was partly due to his extensive (and typical) preparations before the trip.

At the end of his stay, he was invited to write up his recommendations regarding the future of cryogenics at the Tata Institute. Consistent with his belief that cryogenic engineering was an important industrial technology, he suggested that Tata stress applied aspects of cryogenics so as to help India develop a cadre of well-trained cryogenic engineers. These suggestions were accepted and one of the Indian scientists that had helped him in his research was sent to work in Mendelssohn's laboratory in Oxford to learn modern cryogenic practices. [3].

Dr. Mendelssohn summed up his experience at Tata with: "It is refreshing, and almost inconceivable, to find in India an intellectual community whose overriding purpose is the practice of science." [3].

This last quote implies that he did not find all science in India up to the standards he saw at Tata. This was the case and he was quite explicit about it in his report to the Royal Society. During his 4-month stay, he travelled widely in India visiting universities and research establishments as well as cultural sites in places such as Agra and Benares (Varanasi). In general, he was unimpressed by university physics departments in India saying that they "suffer from a lack of funds, a lack of enthusiastic staff and an undue share of intrigue, the latter with a regional bias" [3]. But he was impressed with the National Physical Laboratory in Delhi, tellingly, with its work in developing practical technologies in magnetic materials, duplicating machines and television tubes. This is consistent with his earlier writing that India would be best served by developing their own industries instead of just importing complete factories.

Above all, Dr. Mendelssohn continued to be impressed with the work at the Tata Institute and the various institutes of the Indian Atomic Energy complex. He felt that these institutes were the model for scientific development in India. The institutes were championed by the Indian physicist Dr. H. J. Bhabha who had impressed Mendelssohn not only with his scientific leadership but also with his ability to obtain resources from the Indian government. These efforts were, of course, aided by India's interest in developing both atomic energy and weapons. India tested its first atomic bomb in 1974.

Kurt Mendelssohn summarized his impressions of science in India with: "It seems to me that the performance of the Bhabha Complex has demonstrated the ability of Indians not only to absorb and extend science but also to develop it efficiently and at a level that is close to the Western effort. The question of whether India can expand scientifically and technologically thus becomes largely a political one" [3].

Of all his travels, Mendelssohn's trips to the People's Republic of China are the most fascinating. He was one of the first western physicists to visit China after its founding and he was one of the few westerners to visit during the Cultural Revolution. On these occasions, he had access to high government officials including Mao Tse-tung and Chou En-lai. His China trips directly lead to one of his books; *In China Now* (1969) and indirectly to another *Science and Western Domination* (1976).

China fascinated Kurt Mendelssohn for the same reasons that India did; it was exotic, it possessed an ancient culture and it was in the early stages of developing its scientific and technological infrastructure. During his travels, he was interested not only in the development of scientific research there but also in the daily activities of the people and the cultural and historical monuments. As a guest of Chinese scientific societies, he was able to see a great deal of the country beyond the standard monuments. He made trips to China in 1960, 1962, 1966 and 1971. These trips spanned a wide range of modern Chinese history.

Dr. Mendelssohn's first trip to China in 1960 stemmed from a friendship that he had with Prof. Chou Pei-yuan who visited Oxford from Beijing University in the mid 1940s. Prof Chou had answered Mendelssohn's many questions about China and suggested that a visit might be possible sometime in the future. In 1960, as he was planning a visit to Japan to lecture at universities there, Mendelssohn wrote to Prof. Chou about the possibility of visiting China. Prof Chou was happy to have him come and lecture at Chinese Universities as well as visit research laboratories. This trip and subsequent ones were sponsored by the Scientific and Technological Association of China and by the Chinese Academy of Science (Academica Sinica).

Dr. Mendelssohn was consistently impressed during his trips to China. He was impressed by the friendliness and hard work of the people he met and even those he saw in the street. He appreciated the skill of Chinese scientists and engineers, and particularly with what he saw as the Chinese Government's emphasis on scientific and technological development and its support of scientists and engineers. Consistent with his efforts to move cryogenics out of the laboratory and into industry in the United Kingdom, Dr. Mendelssohn applauded China's emphasis on not just producing more academic scientists but on expanding practical applications: "the emphasis on science as a powerful means for the reconstruction of the country suddenly changed the study of scientific subjects from a somewhat academic pursuit into a most important service to society" [4]. Writing in 1963 after his second trip to China, he describes for a general audience the Chinese approach to technological training. There were basically three tracks for technical education. "For instance, a man who wants to become a petroleum engineer will receive his education at an institute for petroleum technology. Another who wishes to take up chemical engineering in a wider sense, will go to one of the technical universities. Finally, someone who is drawn to the basic study of thermodynamics and would like to devote himself to fundamental work in the field will enter one of the standard types of universities" [4]. He thought that the Chinese government correctly understood that the key to success was training as many people as possible in science and technology.

As a result of the multiple visits, Kurt Mendelssohn was able to discuss the changes that he saw in China and in Chinese science and industry. He compares seeing young girls hand-crafting electrical transformers at an agricultural commune in 1960 with modern factories that he visited in 1962. His observation in 1960 was a result of Mao's "Great Leap Forward" program which encouraged everyone to set up small backyard factories to aid industrialization. This has generally been seen to be a failure. However, Kurt Mendelssohn's view was that low productivity was better than none and that people trained in this hand work would then be the foremen of the new factories. Overall, he thought the pace of industrialization was phenomenal. "The complete change of the technological scene which I witnessed in the interval of only a few years is almost unbelievable" [5]. He also pointed out that basic living standards including the availability of consumer goods had much improved between 1960 and 1967.

It was his visit to China in 1966 that generated the most public interest. It took place during the Chinese Cultural Revolution (1966–1976) a political and social movement started in part by Mao Zedong that called for continuing revolution and radicalism. The Cultural Revolution originated partly as a power struggle among top Chinese leaders but eventually went on to impact all aspects of Chinese life. The Cultural Revolution was marked by a distrust of experts, especially intellectuals, and the development (and manipulation) of grass roots groups known as Red Guards.

Dr. Mendelssohn wrote and spoke a number of times on what he saw of the Cultural Revolution. He described crowds of Red Guards in the streets, public

Fig. 8.3   Kurt and Jutta Mendelssohn with Red Guards after a Discussion Session [7]

readings of Mao's saying on trains and in airplanes and the posting of slogans and suggestions for improvements on public spaces. He engaged some of the Red Guards in a debate about the meaning of the Cultural Revolution and how it would, for example, affect a Tang dynasty love poem. Dr. Mendelssohn described this as "a brisk intellectual exercise, remarkably free from high sounding sentiment or cheap generalities" [6]. See Fig. 8.3. His view was that at worst the Cultural Revolution was essentially benign. He believed that it would not impact China's scientific and technological progress and thought that pronouncements of the Central Committee of the Chinese Communist Party protected scientific workers.

In 1969 Dr. Mendelssohn's book *In China Now* was published. This is mainly a pictorial travelogue of his first three trips to China. There are 196 black and white photos and 69 color photos all taken by him. These pictures cover a wide range of topics including: famous tourist sites, people Mendelssohn met during his travels, street and countryside scenes, photographs of scientific research centers (Fig. 8.4), and equipment and industrial facilities. They were taken throughout China in places such as Beijing, Guangzhou, Xian, Nanjing, Shanghai and Hangzhou. There are photographs from the 1960 trip, in which the young girls building transformers are shown (Fig. 8.5), up through 1966. There are a number of photographs (Fig. 8.6) of Red Guards with Mao's book of sayings, a common sight during the Cultural revolution. The photographs have very simple factual labels. The first section of the book is a list of dates and fuller descriptions of each of the photographs in numerical order.

Looking at these photographs today is just as, if not more, fascinating than it would have been in 1966. The pictures are literally a time capsule showing a country undergoing rapid industrialization but still having a populace that uses draft animals

Fig. 8.4 Zheijiang Technical University [7]

Fig. 8.5 Hand Crafting an
Electrical Transformer [7]

for much transport and does much of its work through manual labor. Due to the explosive growth of the Chinese economy since the 1980s, with the exception of the historical landmarks and some of the scenery, much of what is shown in this book (technology, fashion, transport, and infrastructure) is gone. *In China Now* was interesting in 1969 because it showed a land that few people in the West were able to see. It is interesting now because it shows a land that in many respects no longer exists.

Fig. 8.6 Red Guards [7]

There is one page of photographs that, intentionally or not, is quite ironic. On this page are four photographs. Three of them show sophisticated Chinese made scientific apparatus: an electron microscope, a helium liquefier and a press for making artificial diamonds. The fourth shows one of the mechanical clocks imported into China by the Jesuit missionaries in the seventeenth century to impress the Chinese court with Western technology.

Along with the photographs is a 34-page essay that combines a brief review of China's history and culture along with Kurt Mendelssohn's observations that he made on his trips to China. Here too, he was complimentary. He thought that the agricultural communes were well run and observed that "Up-to-date research at the universities and at technological centers is now backed by an instrument industry which produces many sophisticated items..." [7].

In contrast to India, two aspects of scientific and technological modernization in China stood out as particularly laudatory. First, the Chinese government recognized the value of scientists and engineers and devoted resources to training a large number of them. Also, instead of just importing equipment and factories from the West, China rapidly developed their own industrial capabilities, first by copying Western designs but quickly thereafter by developing their own designs.

The Cultural Revolution is also mentioned in *In China Now*. Dr. Mendelssohn's observations again are essentially positive. All the Red Guards he met were "well-disciplined young men and women who had retained traditional Chinese courtesy." [7] He did say that "possibly there had been some excesses and damage to historical monuments, although we saw no trace of this." [7]. However, Kurt Mendelssohn was in China at the very beginning of the Cultural Revolution, before it started to become really destructive and his observations were quickly made obsolete.

Intellectuals became one of the targeted groups of the Cultural Revolution. Universities were closed. Professors were banished to the countryside to perform menial labor. Large numbers of innocent people were persecuted, sent to prison camps and many were killed. The Cultural Revolution was a disaster that set back Chinese development by a decade. Such effects were evident at the time of the publication (though not at the time of the writing) of *In China Now* (1970).

It is worth noting that as time went on Kurt Mendelssohn did not continue to defend the Cultural Revolution. He did, however, continue to maintain links to Chinese scientists and encouraged China's participation in the international cryogenic community. This is consistent with his lifelong belief in the importance of international scientific cooperation and communication.

In China Now attracted a fair amount of attention and was given mostly positive reviews. The photographs in particular were typically seen as valuable. One review [8] called it "a remarkable pictorial record of the land, the people and their daily life".

It is useful to conclude this chapter by briefly touching on the state of cryogenics in India and China in the twenty-first century. Both countries have robust and modern cryogenic research programs. In the area of university research and training, China and India each have programs in roughly a dozen universities. These programs involve almost all aspects of current cryogenic and superconductivity research. Indian and Chinese researchers and students present papers at international conferences and work as visiting researchers and students in foreign universities and laboratories. Both China and India have built large national facilities using superconducting magnets for fusion energy research and both have built superconducting accelerators for particle physics research. Both countries have also developed cryogenic rocket engines for satellite launchers and space exploration.

China and India have also made significant contributions to large international scientific projects such as the Large Hadron Collider (LHC) in Switzerland and the International Tokamak Experimental Reactor (ITER) in France. India in particular has made or has signed up to make significant contributions to these projects in the area of cryogenics. India provided superconducting magnets for beam correction at the LHC and provided all the cryogenic transfer lines for the ITER project. India is also scheduled to provide superconducting magnets for the Facility for Antimatter and Ion Research (FAIR) under design in Germany. Scientists and engineers trained in China and India play important roles in universities, industry and research institutes throughout the world.

While both countries are still somewhat hampered by underequipped university laboratories and (in the case of India at least) a small indigenous industrial base, they are very competitive with Western countries in the fields of cryogenics and superconductivity. They have come a long way since Kurt Mendelssohn's visits and, given his interest in the expansion of cryogenics in these countries, this would no doubt have pleased him.[1]

---

[1] In both countries, I have met many scientists (including former students) who remember Dr. Mendelssohn fondly.

References

1. K. Mendelssohn, "Moscow Journey", Bulletin of the Institute of Physics, October (1957).
2. K. Mendelssohn, "Science in India", The Listener, September, 1964.
3. K. Mendelssohn, "The Royal Society Leverhulme Visiting Professor to the Tata Institute of Fundamental Research", The Royal Society (1970).
4. K. Mendelssohn, "Aspects of Science in China", Arts & Science (1963).
5. K. Mendelssohn, "Science in China". Nature, Vol. 215 (1967).
6. K. Mendelssohn "China's Cultural Revolution", The Listener, December 1966.
7. K. Mendelssohn In China Now, (Paul Hamlyn 1969).
8. Hereford Evening News, December 4, 1970.

Chapter 9
Writings

"No apology is made for the choice of subject"
K. Mendelssohn, Cryophysics *(1960)*

Kurt Mendelssohn wrote constantly throughout his life. In addition to publishing more than 200 scientific papers and 7 books, he wrote book reviews, letters and opinion pieces in popular media and kept up an extensive correspondence with scientists throughout the world. On top of this were his editorial duties at *Cryogenics*, with the proceedings of the International Cryogenic Engineering Conference and as one of the Editors of the *International Cryogenic Monograph Series* published by Plenum Press. Dr. Mendelssohn kept copies of all his publications and correspondence. These make up the bulk of the Mendelssohn Papers, currently located at the Bodleian Library at the University of Oxford.

It is the story of Dr. Mendelssohn's seven books that is most interesting. Unlike his scientific papers, whose subjects were determined in part by current research needs, funding availability and the interests of his coauthors, the subject of his books were determined by his own interests alone. By looking at their order and subjects, one can trace the arc of Kurt Mendelssohn's intellectual journey. Their number and scope are also one of the things that separate his career from other successful scientists. While he wrote in part to gain additional income, it is clear that he was also excited by these topics and interested in sharing his knowledge of them. Additionally, in writing a series of popular books, he was freed from the stylistic restraints of scientific papers. In these volumes, Kurt Mendelssohn can more easily express personal theories and opinions. In short, he could be as pithy as he wanted and some of his best writing occurs in these books.

His first book, *What is Atomic Energy?* was published in 1946 and resulted from one of his early efforts in the popularization of science. It was part of the Sigma Introduction to Science series published by Interscience Publishers of New York. The series was designed to provide short introductions to the general public on scientific topics. Besides this, Dr. Mendelssohn served as the scientific advisor to the series. His contribution is in part shown by the list of authors of the other books in the series, many of whom he knew personally. These included: John G. Daunt (*Electrons in Action*) who worked with him at Oxford, David Shoenberg

© The Author(s), under exclusive license to Springer Nature Switzerland AG 2021
J. G. Weisend II, G. T. Meaden, *Going for Cold*, Springer Biographies,
https://doi.org/10.1007/978-3-030-61199-6_9

(*Magnetism*) a researcher in cryogenics at Cambridge University[1] and Martin Ruhemann (*Power*) with whom he had worked at Breslau. This ability to find qualified people to contribute to writing projects such as journals, book series and conferences was a consistent hallmark of his career.

The book itself also displayed a number of features that would be seen in his later books. These included the extended use of drawings, diagrams and photographs and the use of a historical approach to describing the topic with an emphasis on the people carrying out the work. Given that the first atomic bombs had just been detonated the previous year, this book was very timely. He points out that due to the destruction of Hiroshima and Nagasaki "Never has scientific achievement been given a worse start in the sphere of human affairs than the release of atomic energy" [1]. Kurt Mendelssohn's stated goal is "To redress the balance between vague uneasiness and scientific truth." [1]. After introductory chapters on the atom and energy, the book goes on to describe radioactivity, quantum mechanics, the structure of the nucleus, fission, the atomic bomb, nuclear reactions in stars and early efforts in nuclear power. He was able to explain these complicated topics clearly without resorting to jargon or mathematical explanations. It is written at a level that non-scientists can understand.

Cryophysics (1960) was Kurt Mendelssohn's second book and his first involving his expertise in cryogenics. Like his first book, this was published by Interscience Publishers of New York as part of a series of short books on physics and astronomy. However, they were written more for science students and Dr. Mendelssohn did not play any other role in the series other than authoring this book. The book is the closest of Kurt Mendelssohn's books to being a textbook. In the introduction, he says that the level of the text is designed for a last year undergraduate or a first-year graduate student. However, even here he points out that the text can "provide a general survey of the field for those interested but not necessarily engaged in it" [2] and is something of a popularization of science.

The tone throughout is professional, detached and expository. The short text of roughly 200 pages uses clear explanations, diagrams and mathematics to describe a variety of basic physical phenomena that occur at low temperatures. Topics covered include: generation of low temperatures, material properties (such as specific heat, thermal and electrical conduction), magnetism, superconductivity and superfluid helium. The book covers a number of research areas that Mendelssohn was person-ally involved in including: the design of small helium liquefiers like those he and Francis Simon built, the study of superfluid helium films and the magnetic cooling techniques using nuclear spins studied by Francis Simon and Nicholas Kurti.

The emphasis of the book is on basic physical phenomena at low temperatures and on how studies at low temperatures help explain these basic phenomena. This emphasis is reflected in the title—*Cryophysics*. It is interesting that, while later in his career he championed practical applications of cryogenics and cryogenic engineer-ing, in his first book on the subject he harkened back to his research experience at the

---

[1]David Shoenberg wrote Mendelssohn's official Biographical Memoir for the Royal Society.

Fig. 9.1 "The Quest for Absolute Zero", book cover (Courtesy T. Meaden)

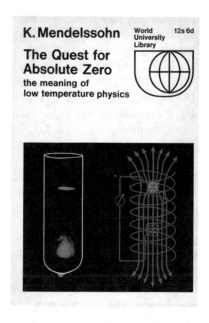

K. Mendelssohn   World University Library   12s 6d

The Quest for Absolute Zero
the meaning of low temperature physics

University of Berlin. There Nernst, Simon, Mendelssohn and others were using cryogenics as a tool to understand the basic physics associated with specific heat and chemical equilibrium. The emphasis in the book on the role of cryogenics in helping to explain basic physical phenomena may, in part, be a response to a fairly recent technological development. Prof. Samuel Collins of the Massachusetts Institute of Technology MIT had developed a helium liquefier in the late 1940s that was commercially produced and could be acquired by university laboratories. This eliminated the need for research groups to build and design their own liquefiers and greatly expanded the number of laboratories able to conduct cryogenic research. Dr. Mendelssohn may have in part meant the book as a response to "Purists who contend that with the advent of the Collins Helium Cryostat low temperature physics as such has ceased to exist. . ." [2].

Due to the book's emphasis on basic physical phenomena as opposed to technology and due to the author's stated goal of ". . . a conscious effort has been made to curb unbalanced enthusiasm for current research" [2], the volume contains little obsolete information and can still 50 years later be read profitably.

Kurt Mendelssohn's next book (see Fig. 9.1), *The Quest for Absolute Zero* (1966), is not only one of his most popular, but also provides a transition between his earlier and later books. Previously, the books were about technical subjects with an eye to popularizing science. After this one, his books concerned the history of science, Egyptology, travel and the spread of innovation. He restricted his writing on technical subjects to scientific papers. To be sure, all of Dr. Mendelssohn's books, including *In China Now* and *The Riddle of the Pyramids,* have at their heart issues of science and technology and their application to other subjects, but his later works cannot be viewed as technical publications or textbooks. The reason for the shift in emphasis may be a desire to reach a broader audience or it may result from simply wanting to describe topics he enjoyed because he always had such broad intellectual interests.

The Quest for Absolute Zero is a history of the development of cryogenics. It is in part a technical book, because he must describe aspects of cryogenics to explain their development, and at the same time it is a history of the events and people associated with the development of cryogenics. The success of this book most likely also made publishers more receptive to Kurt Mendelssohn's later book proposals. In effect, *The Quest for Absolute Zero* enabled the publication of the later, less technical books.

The structure of *The Quest for Absolute Zero* is essentially chronological. This approach fits the development of cryogenics well as there was a race in the late nineteenth and early twentieth centuries among scientists to go to lower and lower temperatures and to liquefy the so-called permanent gases which, at the time, included oxygen, nitrogen, argon, hydrogen and helium. In the first half of his book, Dr. Mendelssohn traces this race from the first liquefaction of oxygen by the French physicist Louis Paul Cailletet in 1877 through James Dewar's liquefaction of hydrogen in 1898 to the liquefaction of helium in 1908 by Kamerlingh Onnes and the subsequent discovery of superconductivity by Onnes in 1911. Along the way, Mendelssohn discusses the people and institutions involved as well as the necessary technical developments such as the cascade cooling technique and the vacuum insulated "Dewar" flask. He has particular praise for Kamerlingh Onnes for his extensive experimental preparations and well-organized laboratory as well as for his development of a scientific journal (*Communications from the Physical Laboratory of the University of Leiden*) describing cryogenic research. Dr. Mendelssohn, of course, shared these interests in proper preparation of experiments and in scientific communication. He also contrasted Leiden's very open approach to the sharing of their techniques and facilities to the more restrictive approach taken by James Dewar at the Royal Institution in London.

The second half of the book is organized thematically into chapters on such topics as the Third Law of Thermodynamics, magnetic cooling, superconductivity and superfluidity. Here too, Dr. Mendelssohn takes an historical approach in describing the developments in each of these areas. His description of the development of the Third Law of Thermodynamics, for example, foreshadows his later book on Walter Nernst. It is worth noting that Mendelssohn himself was directly associated with a lot of the research described in the latter half of the book, including the measurement of the specific heat of materials at cryogenic temperatures and of the behavior of superfluid helium films.

Throughout the volume Mendelssohn shows the linkage between low temperature physics and the development of quantum mechanics and modern physics. The book's subtitle "the meaning of low temperature physics" is a reference to this connection.

While *The Quest for Absolute Zero* is meant for the intelligent layman, Mendelssohn includes simple equations, plots and clear graphical descriptions to explain scientific concepts. He did not in any way talk down to his readers or overly simplify things for them. His use of clear graphical descriptions to explain concepts is a

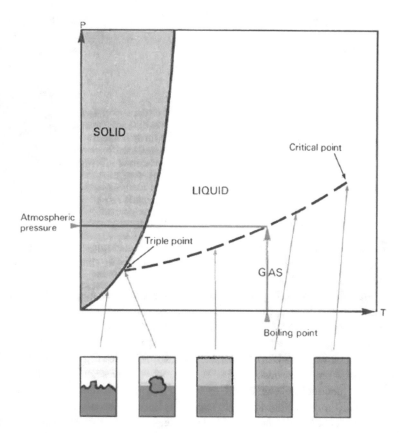

Fig. 9.2 Illustration of a Phase Diagram for Pure Substance from *The Quest for Absolute Zero* [6]

consistent trait of both his scientific and popular writings. An illustration of this is shown in Fig. 9.2. Notice how he links small cartoons representing mixtures of solid, gas and liquid states to their location on the phase diagram, reinforcing the physics of what the diagram represents. Even the distinction between liquid, vapor and super-critical fluid is illustrated.

The Quest for Absolute Zero sold quite well and received a number of good reviews including one from *The Isis* in Oxford which stated that "... the chief features of this book are its clarity and enthusiasm" [3]. The review in *Science* opined that "Mendelssohn has succeeded in conveying to the reader the atmosphere and excitement and mystery which accompanied the unexpected and challenging discoveries near absolute zero." [4].

The book was translated into a number of languages including Spanish, French, German and Japanese. Production of the Japanese edition resulted in a classic Mendelssohn comment. Production had been delayed for a year and Kurt Mendelssohn's publisher (World University Library) ascribed the delay to "usual Oriental slowness" [5]. Dr. Mendelssohn, who knew through his contacts in Japan

Fig. 9.3 Book Cover, *The World of Walther Nernst* (Courtesy T. Meaden)

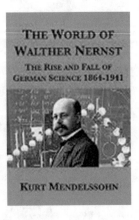

that the delay was principally due to the British publisher, responded "I must assume that Winsley Street is in the Far East" [5], thus demonstrating that he was not going to believe excuses based on national stereotypes.

*The Quest for Absolute Zero i*s still exciting to read and has retained its relevance as one of the better popular histories of cryogenics and scientific discovery.

Kurt Mendelssohn's next book, *In China Now* (1969) has already been discussed in Chap. 8.

His fifth book was *The World of Walter Nernst* (1973) See Fig. 9.3. Whereas *The Quest for Absolute Zero* described at times some of the scientific research that Mendelssohn conducted, *The World of Walter Nernst* is a much more intensely personal book. From the very first line: "So this is Berlin—the city where I was born, had gone to school and studied" [7], it is clear that Kurt Mendelssohn's own experiences would greatly color this book. As a result of this personal connection, the volume contains some of his best writing. Describing the weekly physics colloquium at the University of Berlin that was regularly attended by physicists such as Einstein, Nernst, and Planck (and Mendelssohn sitting in the back of the room as a student): "All that mattered to them was physics, and physics was thrashed out among them, often with the most brutal frankness" [7].

While the book is a biography of the physicist Walter Nernst and uses his life to illustrate the development of German science, it also touches upon a number of other topics. These include: the nature of German universities, the scientific method and the nature of research, the development of the fields of physical chemistry and thermodynamics and the radical transition from classical physics to the concepts of quantum mechanics and relativity. Germany, and in particular Berlin, was a center for much of this work.

The book is also in many ways a nostalgic look at Berlin as a vibrant, intellectual capital that arose at the end of the nineteenth century and died with the rise of the Nazis. This was the Berlin that Kurt Mendelssohn experienced. He makes his views on the 1973 divided city of Berlin quite clear calling the center "dead" and, in referring to the idea that East and West Berlin are separately great cities, says:

"However, nobody is fooled, except perhaps the political pilgrims who come to wail at their respective sides of the wall in tunes of undeserved affluence or insincere class feeling as the case may be." [7].

While the book is relatively short, less than 200 pages, the presence of these secondary topics does not generally detract from its narrative flow. One notable exception might be his lengthy description of German student organizations with their emphasis on the acquisition of fencing scars.

The rise of German science is linked to the rise of Prussia. Prussia's (and later unified Germany's) strong support of universities as well as the importance and resources given to scientists and engineers is seen as the key factor in the rise of German science. Kurt Mendelssohn also points out the practical impact of such support in the resulting world leading chemical and pharmaceutical industries that developed in Germany. The need to properly support science and scientists and the practical economic benefits of doing so is a consistent theme in his writing.

Dr. Mendelssohn describes the fall of German science as a direct result of the rise of the Nazis and their policies of dismissing talented Jewish scientists from their positions; replacing them with second and third-rate talents whose main qualification was their "Aryan" pedigrees. Having experienced this personally, he writes eloquently of the impact these policies had on the many talented scientists who lost their positions and were forced to flee Germany.

These policies, along with a more general anti-intellectualism and the Nazi belief that modern physics was somehow a "Jewish Science" and thus not to be completely trusted, led to the downfall of science in Germany. Dr. Mendelssohn perceptively points out the only real technical area in which Germany bested the allies during World War II was rocketry which "required no new ideas and it had been already pioneered in Germany long before the Nazis came to power" [7]. All other technical innovations important for the war such as RADAR and the development of the atomic bomb came from the allies frequently using the exact same people and principles that the Nazis drove into exile or distrusted.

In the epilogue to *The World of Walther Nernst*, he describes his own journey to Oxford. He is very appreciative of Frederick Lindemann in finding him and his colleagues positions at the University of Oxford. Lindemann, as we have seen was a student of Nernst at Berlin and set up the research program at the Clarendon Laboratory in Oxford as a direct result of his experiences in Berlin. Mendelssohn gives a brief review of Lindemann's accomplishments and dedicates the book in his honor. This linkage between Nernst, Lindemann and Mendelssohn, and between Berlin and Oxford, gives the book a nice symmetry and illustrates the progress of science through personal connections (Fig. 9.4).

As in *The Quest for Absolute Zero,* Kurt Mendelssohn illustrates his story with many anecdotes of the personalities involved. *The World of Walther Nernst* captures a time and place and remains a very good succinct introduction to a number of issues in the history of science.

Kurt Mendelssohn's next book *The Riddle of the Pyramids* (1974) is discussed in the next chapter.

His last published book, *Science and Western Domination* (1976), takes a broader look at the impact of science on the world. The genesis of this book may stem from an earlier book review that he wrote in 1969. This was a review published in *The New Scientist* of *The Grand Titration* by Joseph Needham who, as a Cambridge scholar, was engaged in a lifelong process of investigating and bringing to the attention of the world the great advancements in science and technology found in classical China. By any reasonable analysis, prior to the Scientific Revolution in Europe, China led the world in technological innovation in almost every field. Needham published his findings in an epic (and still growing) series of volumes titled *Science and Civilization in China.* These volumes are a major intellectual landmark in the history of science and technology. *The Grand Titration* is a collection of essays by Needham on science in China and its relationship with the West. In his review, Kurt Mendelssohn was quite positive about the book but did not agree with everything. In particular, he pointed out that China had not always made good use of its developments, nor did these developments necessarily filter out to the west. One of his most telling comments was "Drawing one's own conclusions from Dr. Needham's investigation, one is left paradoxically with awe for that magnificent miracle; the creation of modern science by Western man".

This last comment is consistent with a general question raised by Needham's work. If the Chinese had such a huge advantage in technology over the rest of the world, how is it that China came to be dominated by Western technology and economic power in modern times? The answer for many people, including Kurt Mendelssohn, was the Scientific Revolution that occurred in Europe but not China. This was the subject of *Science and Western Domination*.

One can easily see how such a topic would interest Kurt Mendelssohn. It combined his interests in China (both historical and contemporary) with his interest in the history of science and his strong belief in the importance of science and scientists to civilization in a book that could be aimed at the general population.

Dr. Mendelssohn's thesis in this book is that the reason the West now (i.e. in 1976) dominates the world economically and militarily despite the massive head start that China had in technology was due not just to the Scientific Revolution but more specifically to the rise of the scientific method in the West. His point is that not only does the rigor of the scientific method (observing phenomena, forming hypotheses and testing them) allow us to understand the world in a rational way but it also leads to an efficiency of effort. This efficiency comes about due to the ability of scientists and engineers to accurately predict how systems will respond based on known scientific principles. Moreover, "The value of this prediction lies not so much in allowing us to see into the future but in permitting us to exclude many avenues of progress that have no future" [8]. This results, according to Kurt Mendelssohn, in a more efficient use of intellectual capital. Revisiting his theme of the importance of scientists, Dr. Mendelssohn contrasts the power of the scientific method with politics: "the predictability of political schemes has not appreciatively improved in the last thousand years or so" [8].

Once Mendelssohn has made his point about the strength of the scientific method in the first chapter of the book, what follows is a fairly standard overview of the development of Western science. The usual characters such as Prince Henry the Navigator, Galileo, and Francis Bacon are all present. Dr. Mendelssohn does tell this story in his usual interesting and accessible style and as in his previous books includes intriguing asides about the protagonists. We learn, for instance that Amerigo Vespucci's niece was allegedly the model for the Venus in Botticelli's painting *The Birth of Venus*.

Even in the more standard sections of the book, there are features that reflect Kurt Mendelssohn's interests. There is, for example, an entire chapter discussing the importance of mechanical power production for transportation and manufacturing and the resulting impact on Western dominance. This chapter could almost be subtitled "how thermodynamics conquered the world". There is a similar chapter on electromagnetism and its importance in communication technology. The final chapter of the book discusses the advent of Modern Physics. Certainly, all these points are valid and the chapters correct but it is clear that the book is written from the perspective of a physicist.

Throughout the book, Kurt Mendelssohn makes it clear that he is not claiming that Europeans and North Americans are somehow smarter or more advanced than people in developing countries: "I for one would not hope to win in a mental ability test against a member of the Hanlin Academy or of a Sanskrit college" [8], but rather that the West has dominated in the last 500 years in large part due to its adoption of science. He goes on to say that now countries such as China and India have adopted science they may well catch up with and pass the West in importance.[2] This reflects

---

[2]Today (2020) in the case of China at least, this appears to be true.

his respect for foreign scientists and his strong belief in international scientific communication and cooperation.

With his books, Kurt Mendelssohn appears as a prototype of later, more famous popularizers of science such as Carl Sagan. They share the ability and the interest in explaining complicated subjects to the general public. They are comfortable with using mass media (radio and newspapers for Mendelssohn; television for Sagan) to reach an audience. Like Kurt Mendelssohn, Carl Sagan wrote books on topics outside his specialty of astronomy. Had Kurt Mendelssohn been born somewhat later, a BBC TV series on cryogenics or on the role of physics in history seems almost a certainty.

References

1. K. Mendelssohn, *What is Atomic Energy?*, (Interscience Publishers 1946).
2. K. Mendelssohn, *Cryophysics*, (Interscience Publishers 1960).
3. *The Isis*, November 23, 1966.
4. *Science*, Vol. 155, No. 3768 (1967).
5. K. Mendelssohn Papers, Bodleian Library, University of Oxford.
6. K. Mendelssohn, *The Quest for Absolute Zero*, (Weidenfeld and Nicholson, 1966)
7. K. Mendelssohn, *The World of Walter Nernst*, (University of Pittsburg Press 1973).
8. K, Mendelssohn, *Science and Western Domination*, (Thames and Hud son, 1976).

Chapter 10
Pyramids

"This again is a problem for the slide rule"
K. Mendelssohn: *"A Scientist Looks at the Pyramids".*
American Scientist, *Volume 59 (1971)*

At first glance, Dr. Mendelssohn's book *The Riddle of the Pyramids* (1974), see Fig. 10.1, appears to be the least related to his professional interests.[1] It is in fact quite a stretch from cryogenic engineering to Egyptology. There are, however, a number of reasons that it makes sense for Kurt Mendelssohn to have written this book. In addition to his demonstrated wide range of intellectual interests and his interest in foreign and ancient cultures such as China and India; explaining the reason for the pyramids puts him in a familiar role of trying to understand and explain the unknown. In this case, it is ancient history rather than physics. Moreover, as will be seen, the explanation Dr. Mendelssohn gives for the construction of the pyramids is consistent with his oft stated belief in the value of science and technology and by extension the value of the practitioners of these fields to society.

Recall also, that Kurt Mendelssohn grew up in Berlin at exactly the time in which Egyptology was a significant part of popular culture. Due to excavations in Egypt and subsequent exhibitions in Europe and the USA, motifs from ancient Egypt were used in everything from fashion to Art Deco architecture. German archeologists played a major role in this period and to this day Berlin contains many outstanding artifacts most notably the bust of Nefertiti. It is thus likely that Mendelssohn's interest in ancient Egypt started at an early age.

Dr. Mendelssohn described how he first started to work on this problem in *The Riddle of the Pyramids*. The organizing principle of his book is to take the reader through his thought process in recognizing the questions and proposing the answers. Such an approach, makes it more interesting for the reader. Kurt Mendelssohn makes reference to it as a detective story, but the structure of the book is also meant to

---

[1]*The Riddle of the Pyramids* is directly responsible for this work. As a graduate student, I (J.G. Weisend II) read *The Quest for Absolute Zero* and in the description of the author it noted that Kurt Mendelssohn had written a book on the pyramids. I found it fascinating that a cryogenic engineer would write on such a topic and this biography grew out of that interest.

Fig. 10.1 Book Cover: *The Riddle of the Pyramids* (Courtesy T. Meaden)

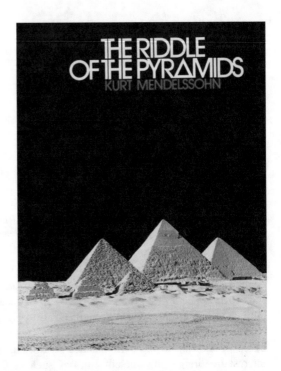

demonstrate that he had no preconceived theories when he started. The ultimate question that Kurt Mendelssohn attempts to answer is "Why were the pyramids built?" but that was not the initial question.

"Like so many detective stories the present one starts with an exotic holiday." [1] So begins Dr. Mendelssohn's description of his interest in pyramids. Returning from a visiting professorship at the University of Kumasi in Ghana in the winter of 1964/1965, he visited Cairo and the pyramids along the Nile. He had previously seen these a few years earlier and was fascinated by the pyramids. Typical of his planning, this time Kurt Mendelssohn had arranged letters of introduction from a Professor of Egyptology at Oxford. These allowed him to see more than the standard tourist sites.

Of particular interest to him was a pyramid at Meidum in which a central core is surrounded by rubble (see Fig. 10.2). The standard explanation was that local people had over the years stripped away the outer surfaces of the pyramid for use in building projects. This explanation struck Mendelssohn at the time as false. Given the size of the pyramid, it did not seem reasonable that such a large amount of material could be moved away and it also appeared that most of the material still existed in the rubble field surrounding the standing core. Kurt Mendelssohn could not explain this mystery at the time but documented it via measurements and photos.

Here we see him applying techniques from science and engineering, such as quantitative measurements and calculations, to a different intellectual discipline. He took the same approach when carrying out work in medicine during the World War II.

Fig. 10.2 The Pyramid at Meidum. Note the rubble surrounding the central core [1]

According to Dr. Mendelssohn, the solution to the Meidum problem came to him in 1966 when a very large pile of tailings from the spoil tip of a coal mine in Aberfan, South Wales suddenly collapsed and slid into the town killing 28 adults and 116 children. Dr. Mendelssohn hypothesized that a similar event had happened in Meidum and that the rubble pile surrounding the core was not due to removal of material by local people but rather was due to the sudden collapse of the outer layers of the pyramid.

In his subsequent writings on the subject, he provides a compelling argument for this explanation. He employs techniques from engineering mechanics such as force balances, slip planes, friction and plastic flow. However, some of the best evidence in support of his theory comes from two other pyramids just north of Meidum.

The "Bent Pyramid" and the "Red Pyramid" located at Dahshur both support Dr. Mendelssohn's theory about the cause of the Meidum pyramid collapse and lead directly to his broader theory about the purpose of the construction of pyramids.

The Bent Pyramid (Fig. 10.3) displays a significant change of slope on its outer surface from 52° to 43.5° at a point where the elevation was about one third complete.

The original explanation for this change was a desire to save building material but Kurt Mendelssohn was able to show by calculation that the actual amount of material saved was not significant. Instead he proposes that this pyramid was being built at the time of the collapse at Meidum and that the builders immediately changed the design at the bent pyramid to prevent a similar failure there. The subsequently-built Red Pyramid starts out with and maintains an outer slope of 43.5° again in theory in response to the collapse at Meidum.

The timing of construction now becomes important and leads Dr. Mendelssohn to his broader and more controversial hypothesis. There is significant evidence (for example unfinished inner stone surfaces in the chambers) that the collapse at Meidum occurred during the final stages of construction. This means that both the Meidum pyramid and the Bent pyramid were under construction at the same time

Fig. 10.3 The Bent
Pyramid [1]

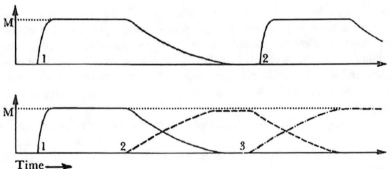

Fig. 10.4 Optimization of the maximum staffing (M) for building pyramids in series (top) or in parallel (bottom) [1]

with the Meidum pyramid much closer to completion. This runs counter to the belief that each ruler built a pyramid to use as his tomb, meaning the pyramids would be built one after the other not at the same time. Dr. Mendelssohn proposed that the pyramids were not built primarily as a tomb for the pharaohs but "On the contrary it seems that they had embarked consciously on the construction of enormous monuments in order to achieve a highly organized political and economic structure of their society" [2]. As part of this theory, he also explains the parallel construction of pyramids as a way to optimize the large labor pool required. In the beginning of construction, large number of people are required while towards the end of construction fewer people are needed. The unneeded people can thus be used to start work on a second pyramid thus keeping them occupied. See Fig. 10.4.

This last point has a definite twentieth century feel to it. In the management of large projects there is always an issue of how to occupy people if the project is delayed. This is literally known as "the standing army problem". Kurt

Mendelssohn's proposed solution for the ancient Egyptians to this problem would be called "resource leveling" today.

The synopsis above does not really do justice to Dr. Mendelssohn's work on pyramids which include detailed discussions of ancient building techniques, descriptions of the improved design features that permitted the Great Pyramids at Giza to be built at the steeper slope of 52° and extends his theories to Central American Pyramids. This is all best understood by reading his writings.

Kurt Mendelssohn put forth his ideas in a number of papers [2–4] and then captured them all in *The Riddle of the Pyramids* published in 1974.

The Riddle of the Pyramids, whose title is a call back to the classical myth of the Riddle of the Sphinx, is a very well written work of popular nonfiction. It is an engaging combination of mystery, travelogue, history and engineering. Everything that Kurt Mendelssohn had learned over the years about describing complicated subjects to the general public is on display here. There are clear diagrams and photographs along with introductory chapters on Egyptology. The text is that of a narrative rather than a textbook or scientific paper which improves its clarity. The book does not drop into jargon or equations to explain concepts.

It is important to keep in mind that this book is not a typical wacky pyramid book. There are no mentions of ancient aliens, pyramid power, the Illuminati etc. Kurt Mendelssohn's hypotheses can all be traced back to observations, calculations and are in many cases supported by well understood engineering concepts.

Upon its release, the book received a number of good reviews. Adrian Berry writing in the *Sunday Telegraph* concluded his review with "this book will make the reader want to go to Egypt and see the pyramids for himself, to clamber over them and peer into their strange chambers. This is just the mental reaction that a good popular scientific book should provoke" [5]. In the journal *Nature*, Cyril Aldred states" "Dr Mendelssohn has a unique contribution to make which deserves very careful consideration by the experts as well as by the general reader..." [6]. A small notice in the *Journal of Archeological Science*, called it a "nicely produced and very readable book" [7].

More than 40 years later, *The Riddle of the Pyramids* still holds up. It remains an interesting, fast-paced book. The reasoning behind Dr. Mendelssohn's explanation of the state of the Meidum pyramid is compelling. His broader theory on why the pyramids were built is plausible but perhaps not yet proven.

Whether Kurt Mendelssohn's theories about the pyramids were correct are not really the most important part of this story. Rather, his sojourn among the pyramids illustrates what is probably the key aspect of his intellectual life; he wanted to use science to find answer to mysteries, whether they occurred in cryogenics or ancient history. Such driving curiosity is a common feature of great scientists. He was a polymath—an all-round scientist who excelled widely in knowledge and learning.

References

1. K. Mendelssohn, *The Riddle of the Pyramids* (Thames and Hudson 1974)
2. K. Mendelssohn, "A Scientist Looks at the Pyramids", *American Scientist*, 59, No. 2, 210 (1971)
3. K. Mendelssohn, "Pyramid Technology", *Bibliotheca Orientalis.*, XXX, No 5/6 (1973)
4. K. Mendelssohn, "A Building Disaster at the Meidum Pyramid", *Jour. Egyptian Arch.* 59, 60, (1973)
5. A. Berry, "Pharaoh's Tactics", *The Sunday Telegraph*, June 2, 1974.
6. C. Aldred, "Physicist on the Pyramids", *Nature*, Vol. 250, July 26, 1974.
7. *Journal of Archaeological Science* (1974).

Correction to: Going for Cold

Correction to:
J. G. Weisend II, G. T. Meaden, *Going for Cold*, **Springer Biographies,**
https://doi.org/10.1007/978-3-030-61199-6

The inadvertently published contents have been corrected as mentioned below.

FM:
In Page XIII, the content "Courtesy M. Mendelssohn" have been removed.

Chapter 1:
In Page 1, the numeral 7 has been removed from the first line, as it has been duplicated.

Chapter 2:
In Page 10, the content ", which was why Simon had given him this as his thesis topic." has been added at the end of the paragraph "The results also cleared up..", below figure 2.2.

Chapter 3:
In Page 14, the word "very" has been added to 3rd para starting with "Franz Simon. . . ."

The updated online version of the book can be found at:
https://doi.org/10.1007/978-3-030-61199-6_1
https://doi.org/10.1007/978-3-030-61199-6_2
https://doi.org/10.1007/978-3-030-61199-6_3
https://doi.org/10.1007/978-3-030-61199-6_4
https://doi.org/10.1007/978-3-030-61199-6_5
https://doi.org/10.1007/978-3-030-61199-6

In Page 16, the caption of fig. 3.3 has been altered as "Kurt Mendelssohn and colleagues in Breslau circa 1931–1933. Top Left—Nicholas Kurti, Middle Left—Rostislav Kaichev, Bottom Row, 3rd and 4th from left—Barbara Zarniko, Kurt Mendelssohn

(Courtesy J. Mendelssohn)

In the paragraph starting with "In 1919...", the content "named to the Professor Lee Chair" has been replaced with "appointed to be the Dr. Lee's Professor"

In Page 17, the content "Mendelssohn's's" in the paragraph starting with "This is....", has been replaced with "Mendelssohn's"

In Page 18, the word "and" has been added between the word "Road" and "was" in the paragraph starting with "Figure 3.4...."

In Page 20, the content "Simon was also being recruited by Cambridge University. Had Franz Simon chosen Cambridge rather than Oxford, Kurt Mendelssohn would have been the senior cryogenics researcher at the Clarendon and his life at Oxford might have been quite different. As it was, Simon was the senior man and was first in line for University Professorships" has been removed from the paragraph starting with "In the fall of 1933...."

Chapter 4:

In Page 24, the caption of fig. 4.1 has been modified as "Fig. 4.1 Kurt Mendelssohn in his Oxford Laboratory circa 1933–1935. Note the open cryostat in the center of the Fig.

In Page 25, the word ", and" has been added next to the content "3.3 K" in the paragraph starting with "Mendelssohn and Moore..."

In Page 39, the word "personal" has been added next to the content "was on" in the paragraph starting with "Kurt Mendelssohn..."

Chapter 5:

In Page 46, the paragraph starting with "After the war ended...." has been removed.

Epilogue:

In Page 102, the word "twentieth" has been corrected next to the content "in the" in the paragraph starting with "Here we are in 2020..."

Epilogue

You cannot become a dedicated man in your field of study unless you know that there is something else in it besides a reasonably good job at the end. For a scientist, this is the sense of adventure in outwitting nature ...

Some Unsolved Problems (The Simon Memorial Lecture 1968):
  K. Mendelssohn, *Cryogenics* June 1968

Kurt Mendelssohn officially retired (Fig. A.1) from the University of Oxford in 1973 and was given the positions of Emeritus Reader and Emeritus Fellow of Wolfson College. He had previously become a Fellow of Wolfson College in 1971. He continued to be active at the start of his retirement, publishing his last book, *Science and Western Domination,* in 1976. He had also started preliminary work on another book that would discuss the links between superfluidity and superconductivity but developed Parkinson's disease from which he died in 1980.

His daughter Monica Jutta Mendelssohn wrote: "Kurt's decline, starting in his mid-sixties, with Parkinson's disease was a great shock to the family, having to follow the sad process of observing a man who had led such an imaginative, creative and energetic life dwindling away and losing all his attractively engaging facial expressions. Jutta made a huge effort in making his later years as interesting and comfortable as possible. She arranged for a cement path with handrail around the garden so that he could independently continue to watch the flowers and shrubs developing throughout the seasonal changes."

In addition to his election as a Fellow of the Royal Society in 1951, Dr. Mendelssohn was awarded the Hughes Medal by the Royal Society in 1967 for his work in superconductivity and superfluidity, particularly for the He II film experiments.

The Institute of Physics presented Mendelssohn with the Simon Memorial Prize in 1968. This prize which is given for excellence in theoretical and experimental low

The original version of the book was revised. The correction to the book is available at https://doi.org/10.1007/978-3-030-61199-6_11

J. G. Weisend II, G. T. Meaden, *Going for Cold*, Springer Biographies,
https://doi.org/10.1007/978-3-030-61199-6

Fig. A.1 Kurt
Mendelssohn in retirement
(courtesy of
M. Mendelssohn)

temperature physics was named in honor of Francis Simon who was Mendelssohn's cousin 12 years older, PhD advisor in Berlin, and colleague at Oxford.

Kurt Mendelssohn's most permanent honor was the creation of an award in his name by the International Cryogenic Engineering Conference (see Chap. 7). Also, at Oxford when the splendid Martin Wood Lecture Theatre was completed at the Clarendon in the year 2000, one of the meeting rooms was named the Mendelssohn Room.

Can his life and his contributions be summarized? Perhaps the best summary can be made (as is frequently true) using his own words. Winners of the Simon Memorial Prize present a lecture as part of the ceremony. Kurt Mendelssohn's lecture in 1968 titled "Some Unsolved Problems" and reprinted in *Cryogenics* illustrates some of the currents of his life. The problems alluded to concern both basic physics (some remaining issues with the Third Law of Thermodynamics and the possible common nature of superfluidity and superconductivity) as well as issues associated with links between academia and industry and the proper organization of universities.

Solving mysteries, whether in physics or other fields and establishing connections that disseminate or increase knowledge, provides a good summary of his life.

Here we are in 2020. Cryogenics continues to grow as a field, particularly in the areas of cryocoolers, large scale helium cryogenics, space cryogenics and liquefied natural gas LNG. Structures developed by Mendelssohn, such as the journal *Cryogenics*, the ICEC and the International Cryogenic Monograph series, remain important aspects of the field. Cryogenics research and industry remain strong worldwide including fast growing and very strong programs in China and India. Not all of this progress is due to Dr. Mendelssohn, but he was there at the beginning of much of it and played an outsized role. He collaborated with and met many of the leading figures in science and politics (Fig. A.2) in the twentieth century.

Unfortunately, also in 2020, there are political and cultural trends that discourage immigration and the creation of links between countries and that are suspicious of expertise and even of curiosity. Here Dr. Mendelssohn's life and career provide an important counterexample. He shows what can be accomplished by refugees, by establishing connections between countries and cultures, by a dedication to excellence and most of all by having a broad curiosity about the world. Perhaps this example is Kurt Mendelssohn's most important contribution.

Fig. A.2 Kurt
Mendelssohn meeting Pope
Paul VI (Courtesy
J. Mendelssohn)

K. Mendelssohn's Publications

Books

1.	*What is Atomic Energy?*
	K. Mendelssohn
	Interscience Publishers 1946
2.	*Cryophysics*
	K. Mendelssohn
	Interscience Publishers 1960
3.	*The Quest for Absolute Zero*
	K. Mendelssohn
	Weidenfeld and Nicholson 1966
4.	*In China Now*
	K. Mendelssohn
	Paul Hamlyn 1969
5.	*The World of Walter Nernst*
	K. Mendelssohn
	The Macmillian Press 1973
6.	*The Riddle of the Pyramids*
	K. Mendelssohn
	Praeger Publishers 1974
7.	*Science and Western Domination*
	K. Mendelssohn
	Thames and Hudson 1973

Papers

1.	Anomale spezifischo Warmon dos festen Wassorataffu boi Helium-temperaturon. F. Simon, K. Mendelssohn and M. Ruhemann. Naturwissonsch. 18, 34, 1930.
2.	Die spezifischon Warmon des feston Wassorstoffe boi Heliumtemperat. K. Mendelssohn, F. Simon and M. Ruhemann. Z. Phys.Chen., 15, 121, 1931.
3.	Eine Apparatur nach dom Descriptionsvorfahren fur Messungen bis zu 2°abs. K. Mendelssohn Z. Physik, 73, 482, 1931.
4.	Uber den Energieinhalt des Bleis in der Nahe des Sprungpunktes der Supraleitfahigkeit. K. Mendelssohn and F. Simon Z. Phys.Chem.B. 16, 291, 1932.
5.	Calorimetrische Untersuchungen im Temperaturgebiet des flussigen Heliums. K. Mendelssohn and John O. Closs. Z. Phys.Chem.B. 19, 291, 1932.
6.	Production of High Magnetic Fields at Low Temperatures. K. Mendelssohn Nature, 132, 602, 1933.
7.	Magnoto-Caloric Effect in Supraconducting Tin. K. Mendelssohn and J.R.Moore. Nature, 133, 413, 1934.
8.	Persistent Currents in Supraconductors. K. Mendelssohn and J.D. Babbitt. Nature, 133, 459, 1934.
9.	Experiments on Supraconductors. T.C. Keeley, K. Mendelssohn and J.R. Moore. Nature, 134, 773, 1934.
10.	Supraconducting Alloys. K. Mendelssohn and J.R.Moore. Nature, 135, 826, 1935.
11.	Magnetic Behaviour of Supraconducting Tin Spheres. K. Mendelssohn and J.D. Babbitt. Proc. Roy.Soc. 151, 316, 1935.
12.	Specific Heat of a Supraconducting Alloy. K. Mendelssohn and J.R.Moore. Proc.Roy.Soc. 151, 334, 1935.
13.	Resistance Thermometry below 10°K. J.D. Babbitt and K. Mendelssohn. Phil. Mag. 20, 1025, 1935.
14.	Discussion on Supraconductivity and other Low Tempera ture Phenomena. K. Mendelssohn. Proc.Roy.Soc. 152, 34, 1935.
15.	Experiments on Supraconductive Tantalum. K. Mendelssohn and J.R.Moore. Phil. Mag. 21, 532, 1936.

(continued)

16.	Magnetic Properties of Supraconductors. T.C. Keeley and K. Mendelssohn. Proc.Roy.Soc. 154, 378, 1936.
17.	Note on Magnetic Hysteresis and Time Effects in Supraconductors. K. Mendelssohn and R.B Pontius. Physica, 3, 327, 1936.
18.	The Transition between the Supraconductive and the Normal State, I—Magnetic Induction in Mercury. K. Mendelssohn. Proc.Roy.Soc. A. 155, 558, 1936.
19.	Time Effects in Supraconductors. K. Mendelssohn and R.B Pontius. Nature, 139, 29, 1936
20.	Absorption of Infra-red Light in Supraconductors. J.G Daunt, T.C. Keeley and K. Mendelssohn. Phil.Mag., 23, 264, 1937.
21.	Supraconductivity of Lanthanum. K. Mendelssohn and J.G Daunt. Nature, 139, 473, 1937
22.	Equilibrium Curve and Entropy Difference between the Supraconductive And the Normal State in Pb, Hg, Sn, Ta and Nb. J.G Daunt and K. Mendelssohn Proc.Roy.Soc. A. 160, 127, 1937.
23.	Supraconductivity. K. Mendelssohn. Phys.Rev. 51. 781, 1937.
24.	Threshold Value, Magnetic Induction and Free Energy of Supraconductors. K. Mendelssohn, J.R. Moore and R.B. Pontius. VII Int. Cong. Froid, The Hague, 1936.
25.	Some Experiments on Supraconductors. K. Mendelssohn, J.G. Daunt and R.B. Pontius. VII Int. Cong. Froid, The Hague, 1936.
26.	Thermal Conductivity of Supraconductors in a Magnetic Field. K. Mendelssohn and R.B Pontius. Phil Mag. 24, 777, 1937.
27.	Thomson Effect of Supraconductive Load. J.G. Daunt and K. Mendelssohn. Nature, 141, 116, 1938.
28.	A Simple Type of Helium Cryostat. J.G. Daunt and K. Mendelssohn. Proc.Phys.Soc. 50, 525, 1938.
29.	Transfer of Helium II on Glass. J.G. Daunt and K. Mendelssohn. Nature, 141, 911, 1938.
30.	Transfer Effect in Liquid Helium II. J.G. Daunt and K. Mendelssohn. Nature, 142, 475, 1938.
31.	Co-existence of Normal and Supraconductive Regions. K. Mendelssohn and J.G. Daunt Phil.Mag. 24, 376, 1938.

(continued)

32.	Photographic Sensitivity and the Reciprocity Law at Low Temperatures. W.F. Berg and K. Mendelssohn. Proc.Roy.Soc. A. 168, 168, 1938.
33.	The Transfer Effect in Liquid He II, I. The Transfer Phenomena. J.G. Daunt and K. Mendelssohn. Proc.Roy.Soc. A. 179, 423, 1939.
34.	The Transfer Effect in Liquid He II, II. Properties of the Transfer Film. J.G. Daunt and K. Mendelssohn. Proc.Roy.Soc. A. 170, 439, 1939.
35.	Surface Transport in Liquid Helium II. J.G. Daunt and K. Mendelssohn Nature, 143, 719, 1939.
36.	Thermodynamical Properties of some Supraconductors. J.G. Daunt, A. Horseman and K. Mendelssohn. Phil. Mag. 27, 754, 1939.
37.	Specific Heat of a Substance showing Spontaneous Electric Polarisation. J. Mendelssohn and K. Mendelssohn. Nature 144, 595, 1939.
38.	The Quantitative Administration of Ether Vapour. R.R. Macintosh and K. Mendelssohn The Lancet, July 19, 1941.
39.	The Oxford Vaporiser No. 1. H.G. Epstein, R.R. Macintoch and K. Mendelssohn. The Lancet, July 19, 1941.
40.	Specific Heat of Supraconductive Tantalum. K. Mendelssohn. Nature, 148, 316, 1941.
41.	The Physics of Blood-Pressure Measurement. D.S.Evans and K. Mendelssohn.
42.	Supraconductivity and Liquid Helium II. J.G. Daunt and K. Mendelssohn Nature, 150, 604, 1942.
43.	The Action of Radiant Heat Cradles. G.M. Brown, D.S. Evans and K. Mendelssohn. Nature, 150, 604, 1942.
44.	A Now Instrument for Visual Determination of Blood-Pressure. Proc.Roy.Soc.Med., July 1942, Vol.XXXVI, No.9 pp 465–467 (section of Anaesthetics, pp. 15–17, 1943. D.S. Evans and K. Mendelssohn.
45.	The Action of Electric Blankets. G.M. Brown and K. Mendelssohn. Br.Med.Journal, Mar., 18, 1944, Vol.i. p.390.
46.	Subcutaneous Temperatures in Moderate Temperature Burns. K. Mendelssohn and R.J. Rossiter. Quarterly Journ. Exp. Physiology & Cognate Med.Sciences, Vol.32,No.4.1944.
47.	Problems of the Transmission of Light Through Tissues and Some Other Media. D.S. Evans (communicated by K. Mendelssohn). Phil.Mag,Ser. 7, Vol.35, p.300, May 1944.

(continued)

48.	Pressure Measurement in Small Anatomical Cavities. D.S. Evans and K. Mendelssohn. The Lancet, April 8, 1944, p.465.
49.	Estimation of Heat Radiation in Clinical Practice. D.S. Evans and K. Mendelssohn. Br.Med.Journ. Dec. 23 1944, Vol. II, p811.
50.	A New Sphygmosope. D.S. Evans and K. Mendelssohn and F. Barnott-Mallison. Br.Journ.Anaesthesia, Vol. XIX, No.3, 1945.
50.a	Transmission of Infection during withdrawal of Blood. K. Mendelssohn and L.J. Witts Br.Med.Journ. Vol.I , p.625. 1945.
51.	The Physical Basis of Radiant Heat Therapy. D.S. Evans and K. Mendelssohn. Proc.Roy.Soc.med. Aug. 1945 Vol.38, No.10,pp 578-586 (section of Physical Medicine, pp 12–20).
52.	The Clinical Application of Heat. D.S. Evans and K.Mendelssohn. Br.Med.Bulletin, Vol.3 (1945) No.6 pp 143–147.
53.	The Frictionless State of Aggregation. K. Mendelssohn. Proc.Phys.Soc. Vol.57, p.371, 1945.
54.	An Experiment on the Mechanism of Supraconductivity. J.G. Daunt and K. Mendelssohn. Proc.Roy.Soc. A. 185, 225, 1945.
55.	Transfer Phenomena and Indeterminacy. J.G. Daunt and K. Mendelssohn. Phys. Rev. 69, NOB. 3 & 4, 126, 1946.
56.	The Accurate determination of limb volume. D.S. Evans and K. Mendelssohn. Brit.Jour.Phys.Med. & Ind.Hyg. 1946.
57.	The Measurement of Infra-Red Radiation for Medical Purposes. D.S. Evans and K. Mendelssohn. Journ.Sc.Instr. 23, No.5, p. 94, 1946.
58.	Zero Point Diffusion in Liquid Helium II. J.G. Daunt and K. Mendelssohn. Nature 157, 839, 1946.
59.	The Visual Measurement of Blood-Pressure. D.S. Evans and K. Mendelssohn. Brit.Med.Bull. 4, No.2, pp. 99–105, 1946.
60.	Conductivity of Sodium-Ammonia Solutions. J.G. Daunt, M. Desirant, and K. Mendelssohn & A.J. Birch. Phys. Rev. 70, Nos. 3 & 4, 219, 1946.
61.	Superconductivity. K. Mendelssohn. Prog.Phys. Vol. X, p.358, 1946.
61.a	Frictionless Transport. K. Mendelssohn. Roy.Coll.Sc.Journ. Vol. XVI, 105, 1946.

(continued)

62.	Conductivity of Sodium-Ammonia Solutions. D.K.C. MacDonald and K. Mendelssohn. & A.J. Birch. Phys.Rev. 71, No.8, 563–564, 1947.
63.	Frictionless Transport. K. Mendelssohn. Phys.Soc.Cam.Conf.Rep. p. 35, 1947.
64.	Experiments on the Formation of the Helium Film. Phys.Soc.Cam.Conf.Rep. p. 41, 1947. By K. Mendelssohn.
65.	A New Technique for Studying the Helium Film. J.B. Brown and K. Mendelssohn. Nature 160, 670, 1947.
66.	Tolerance Limits to Radiant Heat. D.L. Lloyd-Smith and K.Mendelssohn. Brit.Medical.Journ. 1, 975, 1948.
67.	Electrical Resistivity of Alkali Metals Below 20°K. D.K.C. MacDonald and K. Mendelssohn. Nature, 161, 972, 1948.
68.	An Improved Type of Helium Liquefier and Cryostat. J.G. Daunt and K. Mendelssohn. J. Scient. Instrum. 25, 318, 1948
69.	The Superconductive Transition. D.K.C. MacDonald and K. Mendelssohn. Nature, 162, 924, 1948.
69.a.	Viscosity and Superfluidity in Liquid Helium. R. Bowers and K. Mendelssohn. Proc.Phys.Soc., A, LXII, 394, 1949.
70.	Surface Transport of Liquid Helium II. R. Bowers and K. Mendelssohn. Nature, 163, 870, 1949.
71.	Low Temperature Physics. K. Mendelssohn. Rep.Prog.Phys. XII, 270, 1949.
72.	Experiments on the superconductive transition. D.K.C MacDonald and K. Mendelssohn. Proc.Roy.Soc. A, 200, 66, 1949.
73.	Heat Transport in Superconductors. K. Mendelssohn and J.L. Olsen. Proc.Phys.Soc. A, LXIII, p. 2, 1950.
74.	Properties of Superflow in Liquid Helium II. R. Bowers and K. Mendelssohn. Proc.Phys.Soc. A, LXIII, 178,1950.
75.	Resistivity of pure metals at low temperatures. I. I. The alkali metals. II. D.K.C. MacDonald and K. Mendelssohn. III. Proc.Roy.Soc. A, a, 103 1950.

(continued)

76.	Helium II Transfer on Metal Surfaces. K. Mendelssohn and G.K. White. Nature, 166, 27, 1950.
77.	Resistivity of pure metals at low temperatures. II. The alkaline earth metals. IV. D.K.C. MacDonald and K. Mendelssohn. Proc.Roy.Soc. A, 202, 523 1950.
78.	Heat Flow in Superconductive Alloys. K. Mendelssohn and J.L. Olsen. Proc.Phys.Soc. A, LXIII, 1182,1950.
79.	Film Transfer in Helium II: I—The Thermo-Mechanical Effect. J.G. Daunt and K. Mendelssohn. Proc.Phys.Soc. A, LXIII, 1305,1950.
80.	Film Transfer in Helium II: II—Influence of Geometrical Form and Temperature Gradient. J.B. Brown and K. Mendelssohn. Proc.Phys.Soc. A, LXIII, 1312,1950.
81.	Film Transfer in Helium II: III—Influence of Radiation and Impurities. R. Bowers and K. Mendelssohn. Proc.Phys.Soc. A, LXIII, 1318,1950.
82.	Film Transfer in Helium II: IV—The Transfer Rate on Glass and Metals. K. Mendelssohn and G.K. White. Proc.Phys.Soc. A, LXIII, 1328,1950.
83.	Pressure Measurement in Superflow. R. Bowers, B.S. Chandrasekhar, and K. Mendelssohn. Phys. Rev, 80, No. 5, 856–858, 1950.
84.	Anomalous Heat Flow in Superconductors. K. Mendelssohn and J.L. Olsen. Phys. Rev. 80, No. 5, 859–862, 1950.
85.	Superconductivity of Tin Isotopes. W.D. Allen, R.H. Dawton, J.M. Lock, A.B. Pippard, D. Shoenberg, Marianne Bar, K. Mendelssohn and J.L. Olsen. Nature, 166, 1071, 1950.
86.	The viscosity of liquid helium between 2 and 5°K. R. Bowers and K. Mendelssohn. Proc. Roy.Soc. A, 204, P. 365, 1950.
87.	A medium capacity helium liquefier. G.R. Hercus and G.K. White. Journ.SC.Instr. 28, 4–6, 1951.
88.	Heat Conductivity of Liquid Helium I. R. Bowers and K. Mendelssohn. Nature, 167, 111, 1951.
89.	Sub-Critical Flow in the Helium II Film. B.S. Chandrasekhar and K. Mendelssohn. Proc.Phys.Soc. A, LXIV, 512, 1951.

(continued)

90.	Pressure Gradients in Superflow. R. Bowers and G.K. White. Proc.Phys.Soc. A, LXIV, 558, 1951.
91.	The Thermal Conductivity of Cadmium in a Magnetic Field at Low Temperatures. K. Mendelssohn and H.M. Rosenberg. Proc.Phys.Soc. A, LXIV, 1057, 1951.
92.	Experiments on the Unsaturated Helium II Film. R. Boers, D.F. Brewer and K. Mendelssohn. Phil. Mag. Ser.7, Vol. xlii, 1445, 1951.
92.a.	Adiabatic Magnetization of Superconductors. K. Mendelssohn. Nature, 169, 366, 1952.
93.	Helium II Transfer on Metal Surfaces. B.S. Chandrasekhar and K. Mendelssohn. Proc.Phys.Soc. A, LXV, 226, 1952.
94.	Superflow of helium II through narrow slits. R. Bowers and K. Mendelssohn. Proc.Roy.Soc. A, 213, 158, 1952.
94.a.	The Thermal Conductivity of Metals at Low Temperatures. I: The Elements of Groups 1, 2 and 3. K. Mendelssohn and H.M. Rosenberg.
94.b.	The Thermal Conductivity of Metals at Low Temperatures. II. The Transition Elements. K. Mendelssohn and H.M. Rosenberg. Proc.Phys.Soc. A, LXV, 385, 1952.
94.c.	Helium II Transfer Rates on Lucite and Perspex Surfaces. B.S. Chandrasekhar. Phys.Rev. 86, No. 3, 414-415, 1952.
95.	Superconductivity, K. Mendelssohn. Proc. N.B.S. Symposium on Low Temperatures, Washington p.27, 1951.
96.	Flow properties of helium II. K. Mendelssohn. Proc.N.B.S. Symposium on Low Temperatures, Washington p. 165, 1951.
97.	Normal resistivities at low temperatures. K. Mendelssohn. Proc.N.B.S. Symposium on Low Temperatures, Washington p.253, 1951.
98.	Measurements on the Temperature, Current, Magnetic Field Phase Diagram of Superconductivity, K. Mendelssohn, C. Squire, and Tom S. Teasdale Phys.Rev. 87, No. 4,589,1952.
99.	The Formation of Helium Film. D.F. Brewer and K. Mendelssohn. Phil.Mag. 7, 340, 1953.
100.	Anomalous Surface Tension in Helium II. D.F. Brewer and K. Mendelssohn. Phil.Mag. 7, 559, 1953.
101.	Superflow of Helium II through Compressed Powder. B.S. Chandrasekhar and K. Mendelssohn. Proc.Roy.Soc., A, 218, 18, 1953.

(continued)

101. a.	Heat Conductivity of Metals of Low Temperatures. K. Mendelssohn.
101. b.	The Thermal Conductivity of Metals in High Magnetic Fields K. Mendelssohn and H.M. Rosenberg. Proc.Roy.Soc. A, 218, 190, 1953.
101. c.	Inconsistencies in Absorption Experiments of Helium II. D.F.Brewer and K. Mendelssohn. Phil.Mag., 7, vol. 44, 789, 1953.
101. d.	Heat Conductivities of Superconductive Sn, IN, TI, Ta, Cb, and AI below 1°K. K. Mendelssohn and C.A. Renton. Phil.Mag. 7, vol. 44, 776. 1953
101. e.	Thermal Conductivity of Superconductors. K. Mendelssohn. Physics XIX, 775, 1953.
101. f.	A New State of Aggregation. K. Mendelssohn. Royal Inst.Lecture, Nov. 13., 1953.
101. g.	Optical Determination of the Thickness of the Unsaturated Helium II Film. N.C. McCrum and K. Mendelssohn. Phil.Mag., 7, vol.45, 102, 1954.
101. h.	Some Experiments on the Thermal Conductivity of Metals. K. Mendelssohn. Solvay Congress 1954.
101. i.	First Year University Courses. I. K. Mendelssohn. Bull.Inst.P. Sept. 1954.
102.	Der gegenwartige Stand der Tieftemperatphysik. K. Mendelssohn. Physikertagung Hamburg 1954.
102. a.	Heat Conduction in Superconductors. K. Mendelssohn. Progress in Low Temp. Phys. 1955.
103.	The Heat Conductivity of Superconductors below 1°K. K. Mendelssohn and C.A. Renton. Proc.Roy.Soc., A, 230, 157, 1955.
104.	The Entropy of Superfluid Helium. D.F. Brewer, D.O. Edwards, and K. Mendelssohn. Proc.Phys.Soc.A, LXVIII, 939, 1955.
105.	The Heat of Transport of Liquid Helium II. D. Brewer, D. Edwards, and K. Mendelssohn. Conf.de Physique des basses Temperatures, Paris, sept. 1955.
106.	Sub-Critical Flow in the Helium II Film. B.S. Chandrasekhar and K. Mendelssohn. Proc.Phys.Soc. A, LXVIII, 857, 1955.
107.	Adiabatic Magnetisation of Superconducting Tin. K. Mendelssohn and M. Yaqub. Conf. de Physique des basses temperatures, Paris sept. 1955.
108.	Fritz London. K. Mendelssohn. Die Naturwissenschaften, 42, Heft 23 617, 1955.

(continued)

109.	Scattering of Phonons and Electrons by Imperfections in a Metal. K. Mendelssohn and H. Montgomery. Phil.Mag. Aug. 1956.
110.	Electrical and Thermal Conductivity of Metals. K. Mendelssohn. Can. Jour.Phys. 34, 1315, 1956.
111.	The Onset of Friction in Helium II. D.F. Brewer, D.O. Edwards and K. Mendelssohn. Phil.Mag., Dec. 1956.
111.a	Liquid Helium, K. Mendelssohn Handbook Physics 15, 370, 1956.
112.	Low-Temperature Physics in the U.S.S.R. K. Mendelssohn. Nature, 188, 460, 1957.
113.	Moscow Journey. K. Mendelssohn. Bull. Inst P. 336, Oct., 1957.
114.	The Moral for Britain. K. Mendelssohn. New Scientist, Oct. 1957.
115.	Getting to Know Russia's Scientists. K. Mendelssohn. The Daily Telegraph, 5th Feb., 1958.
116.	The Russians value their scientists. K. Mendelssohn. Oxford Mail, 9th Oct. 1957.
117.	The Drive for New Ideas in Physics (A Scientist in Russia – 1) K. Mendelssohn. Observer, Aug. 11th, 1957.
118.	Russia Pays Her Physicists Well. K. Mendelssohn. Observer, Aug. 18th, 1957.
119.	Superfluids. K. Mendelssohn. Science, 127, No. 3292, 215. 1958.
120.	Superconductivity and Superfluidity. K. Mendelssohn. Zeitschrift Physikalische Chemie Neue Folge, 16, 3, 1958.
121.	Liquid Helium. K. Mendelssohn. Handbuch der Physik (Encyclopedia of Physics) XV.
122.	Recent Experiments on Superfluidity. K. Mendelssohn. N.1 Supplemento al Vol 9, Serie X, Nuovo Cimento, 228.
123.	Thermal Conductivity of Superconductors. K. Mendelssohn. Physica XXIV. 1958.

(continued)

124.	Max Planck. K. Mendelssohn. Bull.Inst. P. Jan., 1959.
125.	Fifty Years of Liquid Helium. K. Mendelssohn. Science News 51, Feb., 1959.
126.	Low Temperature Resistivity of Plutonium and Neptunium. J.A. Lee, G.T. Meaden and K. Mendelssohn. Proc.Phys.Soc. LXXIV, 671, 1959.
127.	Science in Poland Today. K. Mendelssohn. New Scientist 24th Dec., 1959.
128.	The Superconduction of Switching and Storage. K. Mendelssohn. Proc. Inst. E.E., 106, B, No 29, Sept. 1959.
129.	Excitations in Liquid Helium. K. Mendelssohn. Symposium van de Koniklijke Vlaamse Academie voor Wetenschappen, 1959.
130.	VI – The Two Liquid Heliums. K. Mendelssohn. M&B Laboratory Bulletin IV No.4.
130a.	Low-Temperature Research on Transuranic Metals. K. Mendelssohn. J – 6.
130b.	The Thermal Conductivity of Metals at Low Temperatures. K. Mendelssohn and H.M. Rosenberg. Solid State Physics. 12 1961.
130c.	Experimental Work on Superconductivity. K. Mendelssohn. IBM Journal, Jan., 1962.
130d.	Accumulation of Radiation Damage in Plutonium. J.A. Lee, K. Mendelssohn, and D.A. Wigley. Cryogenics, March 1962.
130e.	Thermal Conductivity of Tantalum and Niobium below 1°K. A. Connolly and K. Mendelssohn. Proc.Roy, Soc., A, 266, 429, 1962.
130f.	Thermal Conductivity of P-type Geranium between 0.2°K and 4°K. J.A. Carruthers, J.F. Cochran, and K. Mendelssohn. Cryogenics, 2, No.2. 1962.
131.	Effect of Self – Irradiation on the Resistivity of Plutonium. J.A. Lee, K. Mendelssohn, and D.A. Wigley. Physics Letters, Vol. 1, No. 8. 1962.
132.	Science in Communist China. K. Mendelssohn. Cont. Phys., 8, No. 6, 474, 1962.
133.	Accumulation of Radiation Damage in Plutonium. J.A. Lee, K. Mendelssohn, & D.A. Wigley. See 164.

(continued)

133a.	Appearance of Friction in Superfluid Helium. K. Mendelssohn. Rendiconti Scuola Int. Fisica. XXI.
134.	Japanese Science Today. K. Mendelssohn. Listener, Jan. 10th, 1963.
136.	Low Temperature Resistivity of Self-irradiated Plutonium. J.A. Lee, K. Mendelssohn, and D.A. Wigley. Cryogenics, 3, No. 1, 1963.
137.	Accumulation of Self-Damage in Plutonium. J.A. Lee, K. Mendelssohn, and D.A. Wigley. Proc.Int.Conf. on Crystal Lattice Defects, 1962, Conf. J. Phys.Soc. Japan, 18, Suppl. III, 312 (963). 1963.
137a.	Aspects of Science in China. K. Mendelssohn. Arts & Science, p.26, 1963.
139.	Heat Conductivity of Pure Metals below 1°K. G. Davey and K. Mendelssohn. Physics Letters, 7, No. 3, 1963.
140.	Behind the Bamboo Curtain. K. Mendelssohn. B.B.C. Tuesday Talk, Recorded Feb. 8th, 1963.
141.	The Study of Crystal Imperfections in Thermal Conductivity Measurements. K. Mendelssohn. Proc. Int. Conf. on Crystal Lattice Defects 1962. Journ. Phys Soc., Japan 18 Suppl. II, 1963.
159.	Physical Science in the Twentieth Century. K. Mendelssohn. Contemp. Phys. 4, No 6, 433, 1963.
160.	Patterns of Superconductivity. K. Mendelssohn. Cryogenics, Sept. 1963. Vol. 3, No. 3.
162.	Prewar Work on Superconductivity as Seen from Oxford. K. Mendelssohn. Rev. Mod. Phys. 36, No. 1 1964.
163.	Self-irradiation Effects in Stablilized Plutonium. J.A. Lee, E. King, K. Mendelssohn, and D.A. Wigley. Acta Metallurgica, 12, 111, 1964.
164.	On Different Types of Superconductivity. K. Mendelssohn. Rev. Mod. Phys. 36, No. 1, 1964.
167.	Time Effects in Superconductivity. L. Lowell and K. Mendelssohn. Cryogenics, Feb. 1964.
168.	Walther Nernst: An Appreciation. K. Mendelssohn. Cryogenics. 4, No. 3. June 1964.
169.	Onset and Growth of Vorticity in Liquid Helium—II. S.M. Bhagat, P.R. Critchlow, and K. Mendelssohn. Cryogenics, June 1964.

(continued)

170.	The Meaning of Superfluidity. K. Mendelssohn. New Scientist, 23, 772, 1964?
171.	Science in India. K. Mendelssohn. Listener Sept. 24th, 1964.
172.	David Keith Chalmers MacDonald. 1920-1063. Biographical Memoris Roy.Soc. 10, Nov. 1964 K. Mendelssohn.
173.	Superconductivity in High Magnetic Fields. K. Mendelssohn. Bulletin Technique, 62 No. 34.
175.	Two Different Worlds, Soviet Scientist in China. K. Mendelssohn. Discovery XXVI, No5, 1965.
176.	Die Bedeutung des Nernstschen Lebens werkes fur die Tieftenperaturforachung. K. Mendelssohn. Zeitschrift fur Chemie, 5, Heft 6, 201., 1965.
177.	Vorticity in He in Narrow Channels. P.A. Dziwornooh, E.S.R. Gopal, K. Mendelssohn and S. M.A. Tirmizi. Low Temp. Phys. – LT9, A, 1965.
178.	Heat Conductivity of Pure Metals below 1°K. G. Davey, K. Mendelssohn, and J.K.N. Sharma. Low Temp. Phys. – LT9, A, 1965.
179.	Failure of Matthiessen's Rule in Plutonium. E. King, J.A. Lee, K. Mendelssohn, and D.A. Wigley. Low Temp. Phys. LT 9., 1964.
180.	Low Temperature Specific Heat of Plutonium. J.A. Lee, K. Mendelssohn, and P.W. Sutcliffe. Cryogenics 5, 227, 1965.
181.	Plutonium and the Actinide Metals. K. Mendelssohn. Science Journal, Nov. 1965, p. 73.
182.	Resistivity of Plutonium, Neptunium and Uranium due to the Accumulation of Radio-active Self Damage. K. Mendelssohn. Proc.Roy.Soc. A, 284, 325, 1965.
183.	The Effect of Proton Irradiation at Low Temperatures on the Resistance of Manganese. K. Mendelssohn and D.A. Wigley. Phys. Letters, 20 No, 5, 483, 1966.
184.	Scientists in Africa. K. Mendelssohn. Listener, April 14th, 1966.
185.	Science and Technology in China. K. Mendelssohn. Discovery, XXVII, No.6, 8, 1966.
186.	Heat Conductivity of Pure Metals below 1°K. K. Mendelssohn, J.K.N. Sharma, and I. Yoshida. Supplement Bulletin Inst. Int. Froid. 1965 – 2.

(continued)

187.	Dewar at the Royal Institution. K. Mendelssohn. Cont. Phys. 7, No. 5, 331, 1966.
188.	The Clarendon Laboratory, Oxford, (The World of Cryogenics. IV.) K. Mendelssohn. Cryogenics 5, 1–3, June 1962.
189.	China's Cultural Revolution. K. Mendelssohn. Listener, Dec. 8th, 1966.
190.	Dewar at the Royal Institution. K. Mendelssohn. Proc. Roy. Instn 41, No. 189, 1966.
191.	Almost Ideal Behaviour in Some Type II Superconducting Alloys. R.A. French, J.Lowell, and K. Mendelssohn. Cryogenics, April, 1967.
192.	Cashing in on Cryogenics. K. Mendelssohn New Scientist May 25th 1967.
192a.	Science in China. K. Mendelssohn. Nature, 215, No. 5096 10 1967.
193.	Die Geschichte der Supraleitung. K. Mendelssohn. Bild der Wissenschaften, p.219,1967.
194.	Science at the Pyramids. K. Mendelssohn. Science Jou. 48, March, 1966.
196.	Self-Irradiation Damage in the Actinide Metals. K. Mendelssohn. Cryogenics, April 1968.
197.	Some Unsolved Problems. K. Mendelssohn. Cryogenics, 131, June, 1968.
198.	Some Unsolved Problems. K. Mendelssohn. Phys. Bull, 19, 1968.
199.	Die Naturwissenschaften in China. K. Mendelssohn. Unschau, Heft 3, 79, 1968.
200.	Book Review, (The Properties of Liquid and Solid Helium by J. Wilks.). K. Mendelssohn. Cryogenics 8, 36, 1968.
202.	Action of Dislocations on Pinning in a Superconductive Single Crystal. E.L. Andronikashvili, J.G. Chigvinadre, J.S. Tsakadze, R.M. Kerr, J. Lowell and K. Mendelssohn. Phys. Letters A, 28, No. 10, 1969.
203.	Low-Temperature Physics. K. Mendelssohn. Science and Technology, 8, Jan. 1969.

(continued)

204.	Flux Pinning in Thermodynamically Reversible Type II Superconductors. E.L. Andronikashvili, K.G. Chinadze, R.M. Kerr, J.Lowell, K. Mendelssohn, and J.S. Tsakadse. Cryogenics, 119, April 1969.
205.	States of Aggregation. K. Mendelssohn. Phys. Today, 46, April, 1969.
206.	Specific Heat of – Uranium. J.A.Lee, K. Mendelssohn, P.W. Sutcliffe. Phys. Letters, 30, A. No.2, 1969.
207.	Specific Heat of Type II Superconductors. A.A. Melo, and K. Mendelssohn. Phys. Letter. 30, A, No.2. 104, 1969.
208.	Transfer of Helium Film at Low Temperatures. D.J. Martin and K. Mendelssohn. Phys. Letters. 30, A, No.2, 107, 1969.
209.	Cryogenics: The Present and the Future. K. Mendelssohn. Proc. 1st Int. Cryo. Engr. Conf. Tokyo, 1968.
210.	Specific Heats of Plutonium and Neptunium. J.A. Lee, K. Mendelssohn and P.W. Sutcliffe. Proc. Roy.Soc. A, 317, 303, 1970.
211.	A Scientist Looks at the Pyramids. K. Mendelssohn American Scientist, 59, No.2, 210, 1971.
212.	Superfluid ^{4}He Velocities in Narrow Channels between 1.8° and 0.3°K. S.J. Harrison and K. Mendelssohn. Low Temp.Phys. – LT 13 1, 298,
213.	Gedanken eines Naturwissenschaftlers zum Pyramidenbau. Physik in unserer Zeit 3, 40. 1972. K. Mendelssohn.
214.	Superconductivity and Superfluidity. K. Mendelssohn Coopertative Phenomena, 425, 1973.
215.	Pyramid Technology. K. Mendelssohn. Bibliotheca Orientalis., XXX, No 5/6 1973.
216.	A. Building Disaster at the Meidum Pyramid. K. Mendelssohn Jour. Egyptian Arch. 59, 60, 1973.

K. Mendelssohn's Doctoral Students and Their Dissertation Titles

J.D. Babbitt (1934)	Properties of Solids at Low Temperatures.
Judith R. Hull (née Moore) (1937)	Properties of Substances at Low Temperatures Experiments on Superconducting alloys.
John G. Daunt (1937)	Thermal and Magnetic Properties of Substances at very Low Temperatures.
R.B. Pontius (1937)	Magnetic Properties of Superconductors.
D. L. Lloyd-Smith (1947)	The Tolerance of Limits and Physiological Effects of Non-Ionizing Radiations in Man.
J.B. Brown (1948)	Investigation of Phenomena of Frictionless Transport at Low Temperatures.
G.L. Wilson (1949)	Some Physical Properties of Semi-Conductors and Superconductors.
D. Keith C. MacDonald (1949)	Resistivity of Pure Metals at Low Temperatures.
Guy K. White (1950)	Investigations on Liquid Helium.
C.A. Renton (1950)	Some Investigations at Low Temperatures below 1 Absolute.
S.R. Reader (1951)	Temperature Gradient in the Tissues and the Modification of Heat Loss.
Henry M. Whyte (1951)	The Local Thermal Effects of Radiant Heat in Rheumatic and other subjects.
R. Bowers (1951)	Some Properties of Liquid Helium.
J.L. Olsen (1951)	Experiments in Superconductors.
B.S. Chandrasekhar (1952)	Experiments on Transport Phenomena at Low Temperatures.
Harold M. Rosenberg (1953)	Energy Transport at Low Temperatures.
N.G. McCrum (1954)	Experiments in Liquid Helium.
Carl A. Shiffman (1955)	Experiments on Superfluidity.
H. Montgomery (1956)	Thermal and Electrical Properties of Metals at Low Temperatures.
Marianne Olsen-Bär, née Bär. (1956)	Electronic properties of Metals at Low Temperatures.
David O. Edwards (1957)	The Properties of Superfluid Helium.
Peter M. Rowell (1958)	Some Properties of Metals at Low Temperatures.

(continued)

David C. Champeney (1958)	Some Properties of Liquid Helium.
A. Connolly (1960)	Thermal and Electrical Properties of Metals at Very Low Temperatures.
Philip R. Critchlow (1960)	Transport Phenomena in Helium.
G. Terence Meaden (1961)	Low Temperature Properties of Transuranic and Other Metals.
John M. Corsan (1962)	Experiments on the Superconductive Transition
David A. Wigley (1963)	Low Temperature Properties of Transuranic and Other Metals.
Alan F. Rice (1963)	The Heat Conductivity of Metals at Low Temperatures.
J.A. Carruthers (1963/4/5?)	The Thermal Conductivity of p-Type Germanium between $0.2°K$ and $90°K$ and Measurements for Superconductivity in Some Elements below $1°K$.
Richard Burton (1964)	Effect of Structure upon the Superconductive Transition.
Gordon Davey (1965)	Heat Conductivity Experiments below $1°K$ using Helium-3 Cryogenics.
J.K. Sharma (1965)	Thermal Conductivity Measurements at Low Temperatures.
P.A. Dziwornooh (1966)	The Onset of Friction in Helium.
R.A. French (1966)	Some Properties of Superconductors.
J. Lowell (1966)	Some Properties of Type II Semiconductors.
K.V. Rao (1967)	Thermal Conductivity of Metals at Low Temperatures.
Peter M. Sutcliffe (1968)	Low Temperature Properties of Transuranic Metals.
Stephen Blow (1968)	Low Temperature Properties of Actinide Metals.
J. Fairbanks (1968)	The Effect of Treatment on Superconductive Metals.
J.S. Munioz (1968)	Some experiments on the motion of Fluxoids in Type II Superconductors.
Charles S. Griffin (1968)	Self Damage of Actinide Metals.
R. Ratnalingam (1969)	Heat Conductivity Measurements at Low Temperatures.
K.C. Whittaker (1969)	Low Temperature Properties of Manganese and its Alloys.
D.J. Martin (1969)	Properties of Superfluid Helium.
Roger M. Kerr (1969)	Investigations of Type II and High Field Superconductivity.
A.A. Melo (1970)	Specific Heat of Type II Helium Superconductors.
C.A. T. de Sá Furtado (1971)	Losses in Superconductors.
Michael J. Mortimer (1972)	Magnetic and Electronic Properties on Actinide Metals.
M.A.M. de Campos-Tomé (1973)	Thermal Conductivity Measurements at Low Temperatures.

Bibliography

1. D. Brewer, "Kurt Mendelssohn's Research Work", *Proc. ICEC 21*, pp. XXXVIII –XL, Icaris Ltd, Prague (2007).
2. A.J., Croft (Ed), "A Brief History of the Clarendon Laboratory, University of Oxford", Clarendon Laboratory Historical Notes, No 4. (1986).
3. P. Dahl, *Superconductivity*, AIP, New York (1992).
4. N. Kurti, "Franz Eugen Simon" in Biographical Memoirs of the Fellows of the Royal Society, Vol. 4 (1958).
5. G. Klipping, and I. Klipping, "Kurt Mendelssohn and Cryogenic Engineering", *Proc. ICEC 21*, pp. XLI –XLIV, Icaris Ltd, Prague (2007).
6. G. R. Marek, *Gentle Genius: The Story of Felix Mendelssohn*, Funk & Wagnells, New York, (1972).
7. K.D. McRae, *Nuclear Dawn*, Oxford University Press (2014).
8. K. Mendelssohn, "The Future of Cryogenics in Industry", Lecture at the Inaugural Meeting of the British Cryogenics Council (1967).
9. K. Mendelssohn, "The Coming of the Refugee Scientists", The New Scientist, May 26 (1960).
10. K, Mendelssohn, Personal Papers, Special Collections & Western Manuscripts, Bodleian Library, University of Oxford.
11. J. Morell, "Kurt Alfred Georg Mendelssohn (1906 – 1980)", *Oxford Dictionary of National Biography*, Oxford University Press (2004).
12. C. Potak, *Wanderings*, Fawcett, New York (1979).
13. J.H. Sanders, "Nicholas Kurti", Biographical Memoirs of the Fellows of the Royal Society, Vol. 46 (2000).
14. R.G. Scurlock, (Ed), *History & Origins of Cryogenics*, Clarendon Press, Oxford (1992)
15. D. Shoenberg, "Kurt Alfred Georg Mendelssohn", Biographical Memoirs of the Fellows of the Royal Society, Vol. 29 (1983).
16. J.G.Weisend II, *He is for Helium: Defining Cryogenics from ADR to Zero Boiloff*, Cryogenic Society of America (2018).
17. J. G. Weisend II, "Mendelssohn: His Early Years (1906 – 1933)", *Proc. ICEC 21*, pp. XXXIV –XXXVII, Icaris Ltd, Prague (2007).

Personal Communications

Prof. B S Chandrasekhar
Prof. V. Chopa
Prof. N. Kurti
Prof. G. T. Meaden
Dr. J. Mendelssohn
Dr. M. Mendelssohn
Prof. R. G. Scurlock
Prof. K. Timmerhaus

Index